Processing: An Introduction to Programming

Processing: An Introduction to Programming

by
Jeffrey L. Nyhoff
Larry R. Nyhoff

CRC Press
Taylor & Francis Group
Boca Raton London New York

CRC Press is an imprint of the
Taylor & Francis Group, an **informa** business

A CHAPMAN & HALL BOOK

CRC Press
Taylor & Francis Group
6000 Broken Sound Parkway NW, Suite 300
Boca Raton, FL 33487-2742

International Standard Book Number-13: 978-1-4822-5595-9 (Paperback)

Library of Congress Cataloging-in-Publication Data

Names: Nyhoff, Jeffrey, author. | Nyhoff, Larry R., author.
Title: Processing : an introduction to programming / Jeffrey L. Nyhoff, Larry R. Nyhoff.
Description: Boca Raton : CRC Press, 2017.
Identifiers: LCCN 2016041238 | ISBN 9781482255959 (pbk. : alk. paper)
Subjects: LCSH: Processing (Computer program language) | Computer programming--Study and teaching.
Classification: LCC QA76.73.P75 N94 2017 | DDC 005.1071--dc23
LC record available at https://lccn.loc.gov/2016041238

Visit the Taylor & Francis Web site at
http://www.taylorandfrancis.com

and the CRC Press Web site at
http://www.crcpress.com

Printed and bound in the United States of America by
Edwards Brothers Malloy on sustainably sourced paper

To Sharlene, Rebecca, Megan, and SaraKate.

Contents

Chapter 7 ■ Creating void Functions 355

Foreword

In the summer of 2008, I was invited to present at an NSF-sponsored CPATH workshop at La Salle University in Philadelphia. The title of the workshop was *Revitalizing Computer Science Education through the Science of Digital Media*. My first book on Processing had been recently published, and my talk introduced the Processing language and environment. I also discussed my own journey from painter to coder. One workshop attendee seemed especially animated during my talk, and his enthusiasm led to a very lively group discussion. This was my first encounter with Jeff Nyhoff.

Jeff like me is a hybrid, with a background in the arts (theater) and computing. Clearly, what excited Jeff during my talk was discovering another kindred spirit; at the time there were far less of us in academia (*at least out in the open*). Jeff and I are both fascinated by the connections we see between coding and arts practice; this intersection is now commonly referred to as "creative coding." Rather than seeing two radically disparate disciplines—sometimes generalized as extreme opposite, bifurcated left and right brain activities—we see a beautiful integration. But we also understand most (and probably far saner) people won't come to this conclusion on their own. Thus, we both work to spread the creative coding gospel, which has led to our common mission, and the ultimate purpose of this wonderful new book: radically change how computing is taught.

I'm not sure if the workshop was Jeff's initial exposure to Processing, but he instantly recognized its elegance and ultimate usefulness, especially for introductory computing education—something that its originators, Ben Fry and Casey Reas, didn't really consider outside of the arts context. Jeff and I remained in email contact after the workshop, and a few years later, at the annual computer science educators' conference (SIGCSE), we met again in person. It was there, wandering around the SIGCSE book publishers' exhibit, that I met Larry Nyhoff, Jeff's father. Larry was warm, cordial, and humble, and at first I didn't realize Jeff and Larry's professional connection beyond father and son. Through Jeff, Larry knew of me and my work, and too seemed excited by the possibilities Processing offered for the classroom. I left that encounter clueless that I had been speaking to a true computing educator legend.

Learning more about Larry gave me much greater insight into how Jeff, the theater major, became a computer science professor. Larry taught computer science for over 40 years at Calvin College and is the author or co-author of over 30 books, spanning many facets of computer science, as well as related (and seemingly unrelated) disciplines. Larry's books teach numerous programming languages and the aggregate of his publishing efforts literally represents the history of modern computing education. The Processing literature has gotten a lot more respectable thanks to Larry's participation.

Jeff and Larry are deeply committed educators who have been willing to bring a critical, albeit loving, eye to their own discipline, especially around issues of programming pedagogy. I remember receiving many impassioned emails from Jeff discussing his excitement about Processing and CS pedagogy, but also his frustration with the established CS community's initial hesitancy adopting Processing and ultimately embracing change. It was inspiring for me to find someone so committed to helping students, literally *all students*, learn to code. This is why I was especially excited to learn Jeff and Larry were writing their own book on Processing.

The Nyhoffs' book arrives as Processing enters a very stable and mature form. This was certainly not the case for me when I began my book, back in 2004 when Processing was still a messy work in progress. Processing 3 is now a very stable, professional-grade code library and programming environment that is perfectly suited for use within the CS classroom and beyond. Yet, in spite of Processing's success and growing mainstream awareness, there are still very few Processing books designed specifically for the CS classroom. Processing's legacy is the arts and design community, and most of the existing Processing literature supports this population. The book *Creative Coding and Generative Art 2*, of which I am a co-author, was designed for the CS classroom, with a concentration on visual and graphic examples. This approach works well for some populations, but not all. And this is where Jeff and Larry's book comes to the rescue.

The Nyhoffs' new book directly targets the CS classroom in a way that no other Processing book does. The other books, including my own, unabashedly depart from traditional text-based examples found in almost all other introductory programming texts. In our defense, initiating change sometimes requires a small revolution. However, Jeff and Larry present a much less reactionary approach, integrating many of the wonderful things about Processing with traditional approaches that have worked well in CS pedagogy. Not only is their approach sensible and efficient, it's also likely to offer greater comfort to existing CS instructors (who perhaps don't have degrees in theater or painting).

It is this effort of considerate integration—of the *old tried and true* and *new and improved*—that I believe has the greatest chance of tipping the balance for Processing's use in the computing classroom.

Ira Greenberg
Dallas, Texas

Preface: Why We Wrote This Book and For Whom It Is Written

This book is based on our belief that Processing is an excellent language for beginners to learn the fundamentals of computer programming.

What Is Processing?

Processing was originally developed in 2001 at the MIT Media Lab by two graduate students, Ben Fry and Casey Reas, who wanted to make it simpler to use computer programming to produce visual art. Fry and Reas have since been joined by a community of developers who have continued to update and improve Processing. As an open-source software project, Processing is free for anyone to download, and versions are available for Windows, MacOS, and Linux computers.

Because Processing is based on Java, Processing is quite powerful. For example, if you explore such websites as https://processing.org and https://www.openprocessing.org, you'll see a wide variety of works that have been created using Processing: drawings, interactive art, scientific simulations, computer games, music videos, data visualizations, and much more. Some of these creations are remarkably complex.

Because Processing is based on Java, it has many similarities to Java. However, Processing is much easier to learn than Java. In fact, we believe that Processing might well be the *best* language currently available for teaching the basics of computer programming to beginners.

Why Not Java?

One of the most important features of Processing is that it enables us to begin the process of learning to program by building our programs out of simple statements and functions. Unlike strictly object-oriented languages such as Java, Processing does not require us to start with the more complex concepts and structures of "objects." This makes for what we believe is a much gentler introduction to programming. At the same time, because Processing is based on Java, we have the full capability to move on to object-oriented programming whenever we are ready to do so. In fact, this book includes a basic introduction to objects.*

At the same time, because Processing is based on Java, it is usually very similar (and often identical) to Java in terms of much of its syntax and many of its structures, at least in regard to the topics typically covered in an introductory programming course. Thus, we have found that students proceed very smoothly from programming in Processing in a first course to programming in Java in a second course. In general, Processing's syntax and structures carry over readily to the learning of such programming languages as Java, C++, C#, and JavaScript, should you wish to learn any of these programming languages after learning Processing.

What about Python?

Python is becoming increasingly popular as an introductory programming language. However, Python often uses syntax and structures that are somewhat untraditional in comparison with other programming languages that a student might also wish to learn.

We have also found that the dynamic typing of variables in languages like Python is actually easier for students to understand *after* they have learned the static typing of variables in a language like Processing.†

Why This Book?

There are a number of very good books about Processing that have already been written. However, many of them seem to us to be geared first and foremost toward the creation of visual art. Processing certainly provides powerful capabilities for doing this, yet the sophistication of such impressive digital art sometimes comes at the cost of increased complexity in the code, potentially confusing a beginning programmer.

In this book, we focus on using Processing as a language to teach the basics of programming to beginners who may have any of a variety of reasons for wanting to learn to program. Thus, we do not attempt to do a wide survey of Processing's vast capabilities for graphical and interactive works. Rather, we use its capabilities for graphics and interactivity in order to create examples that we hope you will find to be simple, illustrative, interesting, and perhaps even fun. If your goal in learning to program is to create digital art, we believe that the foundation this book provides is an excellent place for you to begin and will prepare you well for other Processing books that are more oriented toward digital art. However, if you are seeking to learn to program for reasons other than the creation of digital art, rest assured that this book has also been written with you in mind.

It might seem surprising to some that a book about such a graphics-oriented programming language as Processing includes many examples that use numbers or text as the primary output. However, these are used for a variety of reasons. First, numerical and textual examples are sometimes easier to understand than graphical ones. Second, most

---

* Chapter 10 introduces objects.
† Processing does have a Python mode, Processing.py, for those who would like to learn to program in Python through Processing. There is also P5.js, a JavaScript "interpretation" of Processing.

graphical programs in Processing have a numerical basis that is often easier to understand in the form of text and numerical output. Third, some readers are likely to be learning to program for the purpose of being able to process numerical information (e.g., if they are interested in math, science, or business) and/or to process data that is likely to include textual information. Fourth, we hope that a combination of simple graphical examples with simple examples that use numbers and/or text will appeal to a broader range of learners. In our own experience, it is sometimes a visual example that first connects with a student, but at other times, it is a numerical and/or textual example that a student finds to be more illustrative. In still other cases, it might be a combination of these types of examples that successfully connects with a student.

We have followed what might be said to be a fairly *traditional* sequence of topics, one that we have found to work well for introducing programming. Also, we hope that such a sequence might seem comfortably familiar to instructors and to students who have undertaken some computer programming before. We have tried to introduce the key computer science concepts associated with introductory programming without going into such detail that risks being overwhelming. Supplementary materials will be available to instructors looking to introduce even more computer science concepts associated with the topics (e.g., in a "CS1" course). Processing's capabilities for animation and interactivity are not explored extensively in this book and are not touched upon until Chapter 7. However, additional examples using animation and interactivity are available for instructors who wish to provide more coverage of these features and introduce them earlier. In addition, several online chapters are provided that introduce slightly more advanced topics in Processing, such as two-dimensional arrays, manipulation of strings, and file input and output (processingnyhoff.com). Additional exercises for most of the chapters are also available.

Throughout the writing of this text, one of our primary concerns has been the *pace* of the material. Some might find it to be slower in comparison with most programming texts. However, in our experience, it can be easy to forget just how gradual a process is required for most beginners to learn to program. Our pace in this book is based on our experiences in teaching programming to a wide variety of beginners in a classroom. No prior programming experience is expected. We hope that you will find the pace of this book to be a good match to the rate at which you learn as well. It is our hope that this text will be useful to students and instructors in a first programming class, but we have tried to write this book in such a way that it will also be useful to persons teaching themselves to program.

Acknowledgments

Many thanks to Randi Cohen at CRC Press/Taylor & Francis Group for her support of this book and her expert guidance throughout the writing process.

Thanks also to Todd Perry at CRC Press/Taylor & Francis Group and Michelle van Kampen and her colleagues at Deanta Publishing Services for their production work on the editing, layout, and typesetting of this text and for their advice during our proofreading.

Thanks to Ira Greenberg for his Foreword and for his encouraging influence. It was one of his demonstrations of Processing that first convinced us of the enormous instructional potential of this programming language.

Thanks to the reviewers, who provided such helpful feedback in response to drafts of this book.

Thanks to faculty colleagues at Trinity for their encouragement and to the many Trinity students who have served as test subjects for most of the material in this book.

Thanks to our families and friends for supporting us in this venture.

Thanks to God for all blessings, including the opportunity to write this book.

Introduction: Welcome to Computer Programming

Congratulations on your decision to try computer programming!

Why Learn to Program?

It may be that programming a computer is something that you aren't sure you are able to do or even want to be able to do. If so, then we hope that this book changes your mind by providing you with a friendly, straightforward, and solid introduction to computer programming.

Being a computer "user" once meant writing programs for the computer. However, the emergence of personal computers beginning in the late 1970s was accompanied by a growth in the availability of "application" software, such as word processors and spreadsheet programs. Such software is written for "end users," who do not need to be computer programmers. Thus, the concept of a computer user as a programmer gradually shifted to that of someone who uses *software written by others*.

However, there are still some very good reasons for learning about computer programming in the current era of computing.

- **Learning computer programming teaches you about what software really is.** Computer software programs may seem complicated and mysterious. However, as you gain experience with simpler examples of computer software, you will start to understand the fundamental nature of computer software. This is important knowledge, given that more and more areas of our lives are affected by computer software.

- **Learning computer programming gives insight into what a programmer does.** Someday, you may find yourself working on a team that includes a computer programmer. Knowing something about what is involved in programming is likely to help you communicate and collaborate better with programmers.

- **Instead of changing to fit the software, you can design how the software works.** When we use software that other people have written, we are often forced to adapt our ways of thinking and working to the way that the software has been designed. In contrast, when you write your own software, you get to decide what the software does and how it works.

- **You cannot always count on someone else to create your software.** Computer software is needed by so many people and in so many ways that there can never be enough people with programming abilities to develop all the software that users would like to have. Thus, knowing enough about computer programming to create or customize software may help you to provide software that you or others need and would not otherwise be available.

- **Even a little knowledge of computer programming is better than none at all.** Many people have no experience at all with programming. Thus, having even a little knowledge of computer programming puts you a step ahead in terms of understanding computing.

- **Computer programming might already be, or turn out to be, something that you need to learn.** Many areas of study and careers are enriched by a basic knowledge of computer programming, and a growing number are starting to require it. Traditionally, programming has paired well with math, science, and business. But in the present era, interests and skills in arts and media also are excellent pairings with computer programming. Today, computer programs are used to solve a very wide range of problems and serve a wide variety of purposes in nearly every area of human society.

- **Computer programming is a creative activity.** Computer programs are often used to solve problems, but computer programming is also a creative activity. Thus, you might find yourself writing a program as a means of expressing yourself or simply out of an urge to create something new. More and more artists and designers are learning to program, especially those who work in electronic media.

- **Computer programming might turn out to be something that you are good at and enjoy doing.** You will never know this without giving programming a try!

These are just a few of the many good reasons for learning about computer programming.

What Is Programming?

Consider a "program" that you might be given at a band, orchestra, or choir concert. This program simply lists the order in which the musical numbers are performed. This notion of putting things in a certain **sequence** is one of the fundamental concepts of computer programming.

Another useful analogy is that of a recipe, where certain actions with various ingredients are performed in a certain order to achieve a certain result. A computer program is much like a recipe, involving the performance of a specific sequence of actions using certain "ingredients" (items of information) for the purpose of producing certain desired results. In computing, such a recipe is also known as an **algorithm**.

A Tour of Processing

We will give a more detailed introduction to Processing in the first chapter, but let's start here with a brief tour of Processing that will give you a little taste of what it's like and where we're headed in this book. *Do not worry* if you do not understand all the details that are presented in this tour; everything in this introduction will be explained more fully in the coming chapters.

Processing is an *open-source* project. It can be downloaded for free from **https://processing.org** and is available for Windows, Mac, and Linux computers.*

When you open Processing, what you see is Processing's integrated development environment or IDE:

An IDE is simply a software environment in which you can develop and test programs. Compared with most IDEs, the Processing IDE is much simpler to use. Officially, the Processing IDE is known as the "Processing development environment" or PDE. Here in the PDE, you can enter programs and have Processing execute them.

"Hello, World!" Example

It is traditional in computer programming to begin with a program that simply displays the words "Hello, World!" Let's create such a program in Processing.

In the Text Editor section of the Processing window,

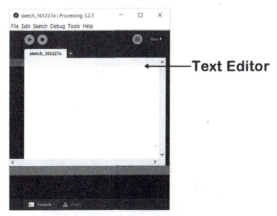

---

* As of this writing, the latest official release is version 3.3.

carefully enter the following:

```
print("Hello, World!");
```

For now, as you type, ignore any error messages that appear on a red bar in the Message Area of the Processing window:

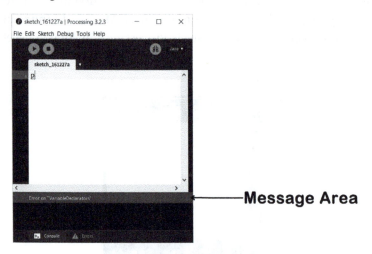—————**Message Area**

After you finish typing, these error messages should disappear. The top portion of the Text Editor area of the Processing window should now resemble the following:

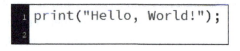

You have now written a line of **code** in Processing. This particular line of code is also an example of what is called a **statement**, because it is a complete instruction to perform a certain action. When we run this program, Processing will carry out this instruction by performing the action it describes.

Next, click the **Run** button, which is found in the area of the Processing window known as the toolbar:

You should see "Hello, World!" displayed in the black section at the bottom of the Processing window. This section of the Processing window is known as the **console**.

If you don't see the above displayed in the console, then here are some things to check in your code:

- The word `print` must be written entirely in all lowercase letters.
- Make sure you have both a left and a right parenthesis.
- The message to be displayed, "Hello, World!" must be enclosed in a pair of double quotes.
- The statement you entered must end with a semicolon.

Once you see the "Hello, World!" greeting, you have successfully written and run your first program! You may now stop the program by pressing the **Stop** button:

Notice the scroll bar on the right edge of the console:

This bar allows you to scroll through the lines of the console output.

There are occasions when it is helpful to enlarge the console area to see more of this program's output. To do this:

1) Click and drag on the edge of the Processing window to enlarge this window.

2) Click and drag upward on the Message Bar to enlarge the console.

You should now be able to view more lines of the console output without having to scroll.

A Processing "Sketch"

Displaying information in the console, such as our "Hello, World!" greeting, can be very useful when programming. However, there is a reason why programs in Processing are also called "sketches": programming in Processing can also involve *drawing*.

As you may have noticed, when you clicked on the Run button, Processing also opened what is called the **display window**. Within this window is a rectangular gray area that we will refer to as the **canvas** because it can be drawn on using Processing functions.

By default, a Processing sketch will render (generate) a canvas that consists of 100 columns of pixels and 100 rows of pixels. A **pixel** is a small square of a single color. In this case, all the pixels that make up the canvas are gray.

The columns of pixels are numbered from left to right, beginning with zero. Likewise, the rows of pixels are numbered from top to bottom, beginning with zero. Thus, each pixel has a unique location, described by the column and the row in which it is located. For example, the pixel in column 50 and row 50 is near the center of the current canvas.

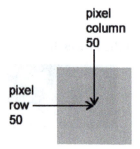

Drawing an Ellipse

Let's delete our current statement that instructed Processing to display "Hello, World!" in the console. The Text Editor area of the Processing window should now be empty:

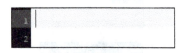

Now, enter the following into the Text Editor area of the Processing window:

```
ellipse(50, 50, 20, 10);
```

This is a statement instructing Processing to draw an ellipse. It includes four items of information that are needed to specify the ellipse that is to be drawn.

The first and second items of information specify where the *center* of the ellipse should be: the pixel column and the pixel row, respectively. In the current statement, we have chosen a pixel location of column 50 and row 50, the center of the canvas:

```
ellipse(50, 50, 20, 10);
```

The third and fourth items specify the *size* of the ellipse to be drawn: the width of the ellipse in pixels and the height of the ellipse in pixels, respectively. In the current statement, we have chosen for the ellipse a width of 20 pixels and a height of 10 pixels:

```
ellipse(50, 50, 20, 10);
```

Notice that we have added a *space* after each comma. This spacing is not required, but this kind of "whitespace" can make code easier to read. In most cases, Processing is very forgiving of adding extra spaces and blank lines for this purpose.

Press the **Run** button.

You should now see an ellipse with its center located at pixel column 50 and pixel row 50, with a width of 20 pixels and a height of 10 pixels:

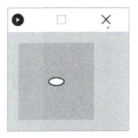

You may now stop the program by pressing the **Stop** button:

Alternatively, you can click to close the display window:

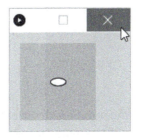

Animation and Interactivity

The preceding ellipse-drawing program gave you a short demonstration of the kinds of programs we will ask Processing to execute. It is an example of a program written in what is called **static mode**. We will use static mode for much of this book because it allows us to write simpler programs.

However, part of what makes Processing powerful and popular is its ability to create animated and interactive drawings. These more advanced capabilities won't be explored extensively in this book and are not introduced until Chapter 7 because they require us to do more advanced programming using Processing's **active mode** and draw() function. However, in the meantime, here is just a little taste of how these are possible in Processing. Again, *do not worry* if you do not understand everything that we are doing. For now, just type in the code and watch what happens.

Start by modifying your program so that it matches the following:

```
void draw()
{
  ellipse(50, 50, 20, 10);
}
```

What we have done here is to enclose our current program inside of what is called the draw() function. What is special about the draw() function is that it automatically "loops." In other words, Processing will automatically perform any statements that are contained within the draw() function over and over again when we run our program.

When we press the **Run** button, we see no change in our sketch:

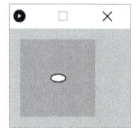

This is because Processing is drawing an ellipse of the same size in the same canvas position, over and over again.

Press the **Stop** button.

Next, let's change the statement that we have placed within the draw() function to the following:

```
void draw()
{
  ellipse(mouseX, mouseY, 20, 10);
}
```

Here, we have replaced the specific numbers describing the pixel column and pixel row for the location of the center of the ellipse with mouseX and mouseY. (Notice that both of these entries end in a capital letter: mouse**X** and mouse**Y**.) These are two "variables" that are built into Processing and continuously keep track of the current pixel location of the mouse pointer when it is over the canvas. (We'll learn much more about variables in the chapters that follow.)

Now, press the **Run** button. If you move your mouse pointer over the canvas, you should see something like the following appear:

We see this result because, over and over again, Processing draws a 20-pixel by 10-pixel ellipse at the current location of the mouse pointer. You now have an *interactive* Processing sketch.

Hopefully, this brief tour gives you a taste of the nature of Processing. Again, this last animated and interactive sketch makes use of some advanced features that won't be covered until Chapter 7, so do not worry about not understanding all the details in this example. Rest assured, the features of Processing will be introduced to you thoroughly and gradually over the course of this book.

Exercises

1) Modify the "Hello, World!" example from this chapter so that when you run the program, your first name is displayed in the greeting instead.

        Example: `Hello, Rebecca!`

2) Modify the code from the "Hello, World!" example in this chapter to the following:

```
println("Hello, World");
println("Welcome to Processing!");
```

Run the modified program to see what output is produced.

3) Modify the previous example so that it produces the following output:

```
Hello, World!
Welcome to Processing!
Let's get started.
```

4) Make the following change to the interactive sample program from this chapter:

```
void draw()
{
   ellipse(mouseX, mouseY, 10, 20);
}
```

Run this program and describe the change in the result.

CHAPTER 1

Basic Drawing in Processing

We'll now put aside the advanced capability associated with the looping `draw()` function and Processing's *active mode* that we explored at the end of the introduction. We will return instead to Processing's simpler *static mode* and work through some simple examples of basic drawing in Processing. These examples will also give us the opportunity to take a closer look at some of the fundamentals of computer programming in general.

This chapter is not an exhaustive introduction to Processing's extensive drawing capabilities. Rather, the goal is to give you enough of a foundation in drawing with Processing to understand the graphical examples in the chapters that follow.

Starting a New Program

In the Processing window, pull down the **File** menu and select **New**.

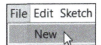

From now on, we will describe such a menu selection as **File > New**.

We now see a new Processing sketch window:

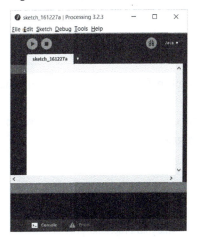

Saving a Program

When starting a new program, it is a good idea to save it promptly and then to save it frequently after making changes to the program. "Save early and often!" is how this strategy is sometimes stated. This is good advice that can help to spare you from losing work that you have done. Saving a program in Processing is very much the same process as saving your work in other software that you use. Select **File > Save**. Since this is the first time you are saving this program, the **"Save sketch folder as…"** dialog box appears.

Save sketch folder as...

Save in:

Using this dialog, start by choosing the file folder location where you would like to save the Processing program that you are currently writing. By default, Processing will save all your sketches (programs) in a file location that is known as your **Sketchbook**. This Processing program setting that specifies this default Sketchbook file location can be changed if you plan to save your Processing programs in a different location.* You can also just browse to the desired file folder location in the manner that you usually do when saving your work in other software programs that you use.

Next, in the **File name** textbox within the dialog box, enter a name for your program. It is customary to capitalize the first letter of a Processing program. Let's name this example **Drawing**.

File name: Drawing

Then, click on the **Save** button to finish saving this program.

Save

Cancel

As a result, Processing creates a new **folder** in which your program is stored. This folder has the same name as that which you gave to your program. Thus, in this example, the new folder is named **Drawing**.

Drawing

Processing calls this the **sketch folder**. If you examine the contents of a sketch folder, you will see a file that has the same name as the folder but also has the **.pde** file name extension. For example, in the current sketch folder, **Drawing.pde** is the name of the file containing our Drawing program.

---

* You can change this default location of the Sketchbook by doing the following:
    Select **File > Preferences**.
    Click the **Browse** button next to the textbox labeled **Sketchbook Location**.
    Browse to the folder where you would like your Processing programs saved.
    Press the **Open** button.
    Press the **OK** button. The folder you selected is now the default location where Processing will save and retrieve your programs.

If you do *not* see the .pde portion of this file name, then you will need to turn on the viewing of file name extensions in your computer's operating system settings if you wish to see such file name extensions.*

Each .pde file created by Processing is simply a text file that stores all the lines of code that we write for a particular program.

Notice that once we save our program, the name we gave this program now appears both at the top of the Processing window and on a tab just above the Text Editor area:

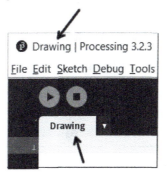

For now, close Processing by selecting **File > Quit**.

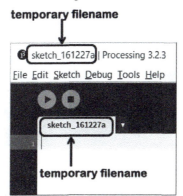

Retrieving a Program

Restart Processing.

Notice that Processing always starts with a *new* program. Processing indicates that this new program has not yet been saved with a name by displaying a temporary name (based on the current date) in both the Processing window and the tab just above the editing area:

Here's a reliable way to retrieve a program that you have previously written and saved in Processing. Select **File > Open**. In the dialog that appears, locate the **Drawing** folder. (If

---

* For example, in Windows 10, simply open the File Explorer, click the View tab, and check the box labeled "File name extensions." For a Mac running OS X, open Finder > Preferences and then select "Show all filename extensions."

you saved your program in a location other than the default location, browse to that other location.)

Double-click the Drawing folder to open it. You should now see the file named **Drawing.pde** that contains your program. (If you do not have file extensions turned on, then just Drawing will be shown as the name of this file.) Select this file.

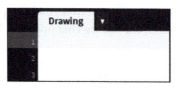

Next, press the **Open** button.

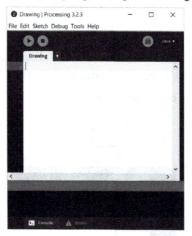

You should now see your Drawing program open once again in Processing:

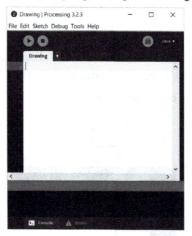

Entering Code into the Text Editor

Our Drawing program does not yet contain any lines of code:

However, Processing still allows us to run this program. We can run a program by pressing the **Run** button.

Alternatively, we can select **File > Run**.*

When we run our drawing program, we once again see the default canvas that Processing generates inside the display window:

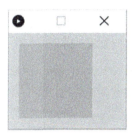

This canvas is a rectangular area that, by default, is comprised of columns and rows of gray pixels.

The word *pixel* is short for "picture element." This term reflects the fact that a pixel is simply a single-colored square that is but one piece of a larger image.†

As we learned previously, the **default width** of the Processing canvas is **100** columns of pixels:

(*Pixels* is sometimes abbreviated *px*.) These pixel columns are numbered from left to right, starting at 0. Thus, for the default canvas, the pixel column numbers run from 0 to 99:

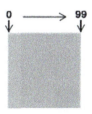

Likewise, the **default height** of the canvas is **100** rows of pixels.

---

* The file menu also shows a keyboard shortcut for running the program: Ctrl-R for Windows, Command-R (⌘-R) for Mac.

† This could be a piece of an image on the screen, a printed image, or an image in a file. The term *pixel* is used in all these contexts, which makes it a somewhat tricky term at times.

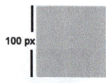

These rows of pixels are numbered from top to bottom, starting at 0. Thus, for the default canvas, the pixel row numbers also run from 0 to 99:

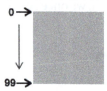

Accordingly, each pixel can be said to have a unique location, described by the column and row in which it is found. For example, the pixel near the center of the default canvas would be the pixel at column 50 and row 50:

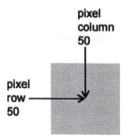

Processing has many built-in **functions** that are available for us to use. For example, during our "tour" of Processing, we used two such functions:

`print()`   displays text in the console
`ellipse()`   draws an ellipse on the canvas

In addition, the **size()** function lets us set the size of the canvas we are using. For example, let's enter the following statement into our program:

```
size(150, 200);
```

Notice that as you type, the area of the Processing window known as the Message Area display turns red and displays error messages:

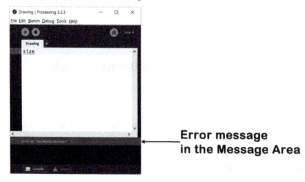

Do not be alarmed by this!

Also, as you type, certain portions of what you have entered may be underlined with a red squiggly line. Again, do not be alarmed by this.

underlining

After you have finished typing the line of code, the error messages should disappear from the Message Area:

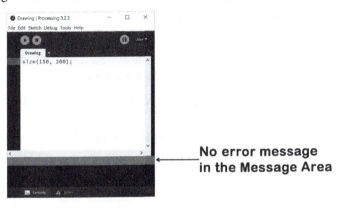

After you have finished typing the line of code, the error messages should disappear from the Message Area.

No error message in the Message Area

Also, the red squiggly line should disappear from beneath your line of code:

size(150, 200); ← **no underlining**

Such error messages and underlining by Processing might be a little confusing at first. However, while you are beginning to learn to program, it is a good habit simply to ignore these warnings from Processing until you have *completely entered* your line of code. If you have correctly entered your line of code, the warning messages and squiggly lines will disappear once you have finished entering it.

Once you become accustomed to these warnings, you will find that they actually can be quite useful. For example, try deleting the semicolon at the end of the line of code you just entered:

size(150, 200)

delete the semicolon

Notice that, now, a red squiggly line has appeared beneath the place in our line of code where we had previously entered a semicolon:

size(150, 200)

Notice also that an error message has once again appeared in the Message Area of the Processing window:

Missing a semicolon ";"

(Note: If you don't see this error message in the Message Area, then *click* on the squiggly red line with your mouse, and the error message will appear.) This is a useful error message. Most statements in Processing must end with a semicolon, and Processing is alerting us here that a semicolon is indeed required in this case. Processing is also warning us that if we try to run our program in its current state, we will receive an error. Most of Processing's error messages give us useful information like this.

As you program with Processing, you will occasionally encounter an error message that isn't entirely clear. However, one of the great improvements in recent versions of Processing is that the error messages have become much easier for beginning programmers to understand.

As soon as we correct our error by reinserting the semicolon

```
size(150, 200);
```
↑
add the semicolon

the squiggly line once again disappears, and the error message disappears from the Message Area.

Resave this `Drawing` program by selecting **File** > **Save**. We will continue working with this program in the next section.

Basic Drawing with Graphical Elements

Setting the "Canvas" Size: A Closer Look at the `size()` Function

When we run our `Drawing` program again, we now see the following display window:

Within the display window is a canvas that is 150 pixel columns in width and 200 pixel rows in height.

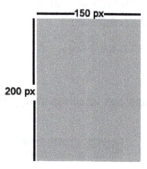

Let's take a closer look at our use of the size() function. When we use a function in a statement like this, we say that we **call** this function. Like most functions, the size() function requires some information from us in order for it to be able to operate when we call it. We put such required information inside the parentheses that follow the name of the function. Such a required item of information is known as a **parameter**. The item supplied by us to fulfill this requirement is known as an **argument**. In other words, for each required *parameter*, we need to supply a corresponding *argument*.

The size() function has *two* required *parameters*, so we need to supply *two* corresponding *arguments*: the desired pixel *width* and the desired pixel *height*, listed in that order, separated by commas.

Thus, the basic form of a call to the size() function could be described in this way:

```
size(width, height);
```

The italicized words are the required parameters:

width	The desired pixel width of the canvas
height	The desired pixel height of the canvas

The word *syntax* is sometimes used in computer programming to describe the correct order and structure of the particular components that make up a larger element of the programming language. For example, the basic form of a call to the size() function is also known as that function's syntax.

In our current program, we have supplied the following arguments in our call to the size() function, specifying the desired canvas *width* and canvas *height*, respectively:

```
size(150, 200);
```

It is important to notice that the *order* in which we supply these arguments is significant. For example, if we had reversed the numbers

```
size(200, 150);
```

then the specified canvas *width* would have been 200 columns of pixels, and the specified canvas *height* would have been 150 rows of pixels. These two arguments to the size() function would have specified a canvas with a different size:

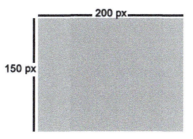

Thus, make sure that the arguments in your call to the size() function appear in the correct order:

```
size(150, 200);
```

Select **File > Save** to resave the current program. We will continue working with this program in the next section.

Drawing Points: The point() Function

We can draw a "point" at a specific pixel location using the point() function. For example, let's add the following statement to our program:

```
size(150, 200);
point(75, 100);
```

When we run our program and look carefully, we see a single black pixel near the center of our canvas:

As we have seen, by default, Processing draws a point on the canvas as a single black pixel. Why was the point draw at this particular canvas location? Because the first argument we supply to the point() function

```
point(75, 100);
```

specifies the pixel *column* where the point is to be located. The second argument we supply to the point() function

```
point(75, 100);
```

specifies the pixel *row* where the point is to be located.

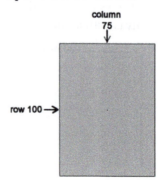

In summary, the basic form (syntax) of a statement making a call to the point() function is

```
point(column, row);
```

The parameters are

> *column*   The pixel column where the point is to be drawn
> *row*   The pixel row where the point is to be drawn

Our program is currently comprised of two statements:

```
size(150, 200);
point(75, 100);
```

It is very important to understand that when we run our program these statements are executed (performed) *by Processing in the order in which they are listed*:

1) size(150, 200); Generates a 150-pixel by 200-pixel blank canvas

2) point(75, 100); Draws a point at pixel column 75, pixel row 100

For example, if we changed the order of these statements to

```
point(75, 100);
size(150, 200);
```

then the order of execution would be

1) point(75, 100); Draws a point at pixel column 75, pixel row 100 on the *default* 100-pixel by 100-pixel canvas

2) size(150, 200); Generates a 150-pixel by 200-pixel blank canvas, *erasing* the point that was drawn

As a result, when we ran the program, we would see a blank canvas.

Thus, make sure your two statements in your program appear in the following order:

```
size(150, 200);
point(75, 100);
```

Select **File > Save** to resave this program. We will continue working with it in the next section.

Drawing Line Segments: The line() Function

We can also draw line segments in Processing using the **line()** function. For example, let's change the second statement in our program to the following:

```
size(150, 200);
```

```
line(0, 0, 75, 100);
```

Now, when we run our program, we see the following:

This particular line segment is drawn because the *first* and *second* arguments we supply to the line() function

```
line(0, 0, 75, 100);
```

specify the column and row, respectively, of the pixel location of one endpoint of the line segment:

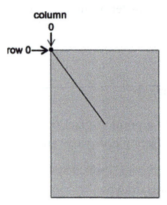

Likewise, the *third* and *fourth* arguments we supply to the line() function specify the column and row, respectively, of the pixel location of the *other* endpoint of the line segment:

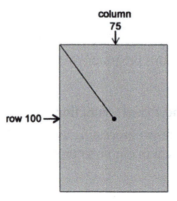

In summary, the basic form of a call to the line() function is

```
line(column₁, row₁, column₂, row₂);
```

The parameters are

column₁	The pixel column location of the first endpoint
row₁	The pixel row location of the first endpoint
column₂	The pixel column location of the second endpoint
row₂	The pixel row location of the second endpoint

Thus, we simply supply the pixel locations of the endpoints of the line segment and Processing does the rest. It determines which pixels need to be recolored in order to generate the line segment we specify in the arguments of our call to the line() function. (By default, this is a line segment that is black and is one pixel in thickness.) This process by which Processing figures out the colors needed for each pixel of the canvas is sometimes known as **rendering**.

Drawing Rectangles: The rect() Function

Processing provides us with the ability to draw a rectangle on the canvas using the **rect()** function. For example, let's change the second statement in our program to the following:

```
size(150, 200);
rect(0, 0, 75, 100);
```

Now, when we run our program, we see the following:

Why does Processing render this particular rectangle? It's because the *first* and *second* arguments that we supply to the rect() function

```
rect(0, 0, 75, 100);
```

specify the *column* and *row*, respectively, of the pixel location of the *upper-left corner* of the rectangle:

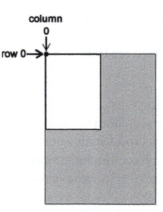

In addition, the *third* and *fourth* arguments that we supply to the rect() function

```
rect(0, 0, 75, 100);
```

specify the rectangle's pixel *width* and pixel *height*, respectively:

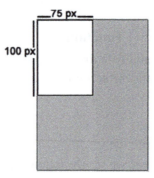

Thus, in general terms, we describe a rectangle with the rect() function by specifying the pixel location of its upper-left corner, its width, and its height:

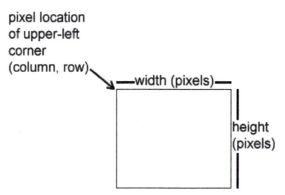

For example, if we change the *first* and *second* arguments in our call to the rect() function

```
size(150, 200);
rect(50, 60, 75, 100);
```

then, when we run our program, we see a rectangle that is the same size as before

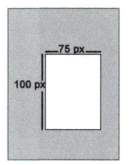

but its upper-left corner is now at pixel column 50 and pixel row 60:

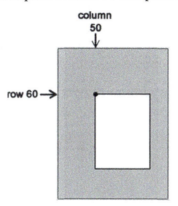

Likewise, if we change the *third* and *fourth* arguments in our call to the rect() function to the following,

```
size(150, 200);
rect(50, 60, 90, 75);
```

then, when we run our program, we still see a rectangle that has its upper-left corner at pixel column 50 and pixel row 60:

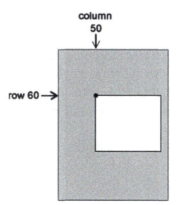

However, the rectangle's *width* is now 90 pixels, and its *height* is now 75 pixels:

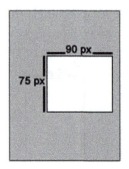

Processing's rect() function also has an optional *fifth* parameter for drawing a *rounded* rectangle. For example, if we supply a fifth argument for an amount of rounding we would like to have applied to each corner of the rectangle,

```
size(150, 200);
rect(50, 60, 90, 75, 20);
```

then, when we run our program, we instead see a rounded rectangle:

In summary, the basic form of a call to the **rect()** function is

```
rect(column, row, width, height [, rounding] );
```

The parameters are

column	The pixel column location of the rectangle's upper-left corner
row	The pixel row location of the rectangle's upper-left corner
width	The width of the rectangle in pixels
height	The height of the rectangle in pixels
rounding	The amount of rounding of the corners (optional)

Thus, with these simple items of information from us, Processing is able to render a rectangle.

Notice that, by default, the edge of the rectangle is *black*. The color used for drawing points, lines, and the edges of shapes is known as the **stroke color**. Notice also that, by default, the rectangle is filled with *white* pixels. The color used to fill shapes is known as the **fill color**. We will learn later in this chapter how to change both the stroke color and fill color.

There is not a function in Processing for drawing a *square*. However, we can easily draw a square using the `rect()` function if we simply specify the rectangle's width and height to be the same number of pixels:

```
size(150, 200);
rect(50, 60, 80, 80);
```

Select **File > Save** to resave this program. We will continue working with it in the next section.

Drawing Ellipses: The `ellipse()` Function

Processing also makes it easy for us to draw ellipses using the **`ellipse()`** function. For example, let's change the second statement in our program to the following:

```
size(150, 200);
ellipse(50, 60, 90, 75);
```

Now, when we run our program, Processing draws the following:

We see this particular ellipse because the *first* and *second* arguments we supply to the `ellipse()` function

```
ellipse(50, 60, 90, 75);
```

specify the *column* and *row*, respectively, of the pixel location of the *center* of the ellipse:

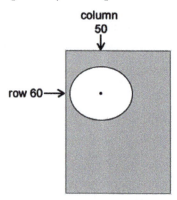

In addition, the *third* and *fourth* arguments that we supply to the `ellipse()` function

```
ellipse(50, 60, 90, 75);
```

specify the ellipse's pixel *width* and pixel *height*, respectively:

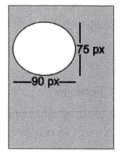

In summary, we describe an ellipse with the `ellipse()` function by specifying the pixel location of its center, its width, and its height:

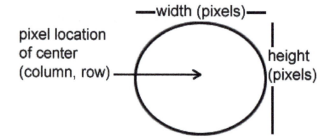

Thus, the basic form of a call to the **ellipse()** function is

```
ellipse(column, row, width, height);
```

The parameters are

column	The pixel column location of the ellipse's center
row	The pixel row location of the ellipse's center

> width   The width of the ellipse in pixels
> height   The height of the ellipse in pixels

With these four simple items of information from us, Processing is able to render an ellipse.

Now, let's try revising the second statement of our program so that we put the location of the center of the ellipse at pixel column 0 and pixel row 0:

```
size(150, 200);
ellipse(0, 0, 90, 75);
```

When we run our program, we now see the following:

We see this result because the center of our 90-pixel by 75-pixel ellipse is now located at pixel column 0 and pixel row 0, the pixel in the upper-left corner of the canvas:

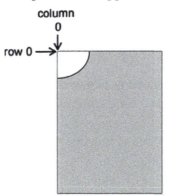

Thus, most of the ellipse we have specified essentially "spills" off the edge of the canvas.

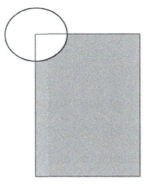

Notice that Processing does not give us an error message when we instruct it to draw points, lines, or shapes that extend beyond the edges of the canvas. This provides a considerable degree of convenience and flexibility as we experiment with drawing. We are even permitted to use *negative* numbers when we wish to specify pixel columns and/or pixel rows that are outside the visible canvas.

Note that there is not a function in Processing for drawing a *circle*. However, we can easily draw a circle using the ellipse() function simply by specifying the ellipse's width and height to be the same number of pixels. For example, the following program draws a circle with a diameter of 50 pixels in the center of the canvas:

```
size(150, 200);
ellipse(75, 100, 50, 50);
```

Select **File > Save** to save this program. We will continue working with it in the next section.

Drawing Triangles: The triangle() Function

Let's learn about another basic shape function that Processing provides: the **triangle()** function.

The triangle() function allows us to draw a triangle simply by specifying *three* pixel locations on the canvas, each one corresponding to one of the three vertices (corner points) of that triangle.

For example, let's change the second line of our Drawing program to the following:

```
size(150, 200);
triangle(0, 0, 0, 60, 90, 75);
```

When we run our program, we see the following:

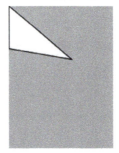

Processing draws this particular triangle because the *first* and *second* arguments we supply to the triangle() function

```
triangle(0, 0, 0, 60, 90, 75);
```

specify the *column* and *row*, respectively, of the *first* vertex (corner point) of the triangle:

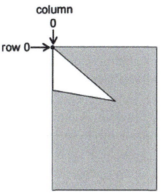

Likewise, the *third* and *fourth* arguments we supply to the `triangle()` function

```
triangle(0, 0, 0, 60, 90, 75);
```

specify the *column* and *row*, respectively, of the *second* vertex of the triangle:

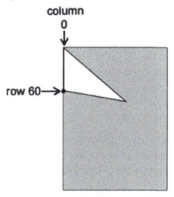

And, lastly, the *fifth* and *sixth* arguments we supply to the `triangle()` function

```
triangle(0, 0, 0, 60, 90, 75);
```

specify the *column* and *row*, respectively, of the *third* vertex of the triangle:

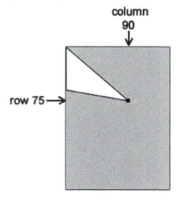

In summary, the basic form of the **triangle()** function is

$$\texttt{triangle}(column_1,\ row_1,\ column_2,\ row_2,\ column_3,\ row_3);$$

The parameters are

$column_1$	The pixel column location of the first vertex
row_1	The pixel row location of the first vertex
$column_2$	The pixel column location of the second vertex
row_2	The pixel row location of the second vertex
$column_3$	The pixel column location of the third vertex
row_3	The pixel row location of the third vertex

Note that the *same triangle* will be drawn regardless of the *order* in which the pixel locations of the three vertices are listed. In other words, it does not matter which of the three pixel locations is considered to be the first, second, or third vertex:

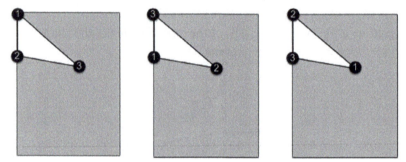

Select **File > Save** to save this program. We will continue working with it in the next section.

Drawing Quadrilaterals: The quad() Function

Processing's **quad()** function allows us to draw a quadrilateral simply by specifying *four* pixel locations on the canvas.

For example, if we modify the second statement of our current program to

```
size(150, 200);
quad(0, 0, 0, 60, 90, 75, 110, 35);
```

then, when we run our program, we see

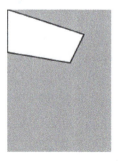

Processing draws this particular quadrilateral because the *first* and *second* arguments we supply to the quad() function

```
quad(0, 0, 0, 60, 90, 75, 110, 35);
```

specify the *column* and *row*, respectively, of the pixel location for the *first* vertex of the quadrilateral:

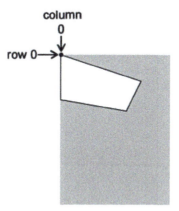

The *third* and *fourth* arguments we supply to the quad() function

```
quad(0, 0, 0, 60, 90, 75, 110, 35);
```

specify the *column* and *row*, respectively, of the pixel location for the *second* vertex of the quadrilateral:

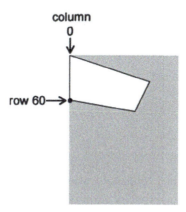

The *fifth* and *sixth* arguments we supply to the quad() function

```
quad(0, 0, 0, 60, 90, 75, 110, 35);
```

specify the *column* and *row*, respectively, of the pixel location for the *third* vertex of the quadrilateral:

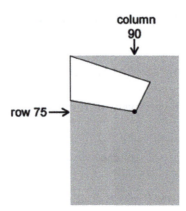

The *seventh* and *eighth* arguments we supply to the quad() function

```
quad(0, 0, 0, 60, 90, 75, 110, 35);
```

specify the *column* and *row*, respectively, of the pixel location for the *fourth* vertex of the quadrilateral:

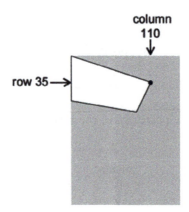

In summary, the basic form of the quad() function is

```
quad(column₁, row₁, column₂, row₂, column₃, row₃, column₄, row₄);
```

The parameters are

$column_1$	The pixel column location of the first vertex
row_1	The pixel row location of the first vertex
$column_2$	The pixel column location of the second vertex
row_2	The pixel row location of the second vertex
$column_3$	The pixel column location of the third vertex
row_3	The pixel row location of the third vertex
$column_4$	The pixel column location of the fourth vertex
row_4	The pixel row location of the fourth vertex

In the case of the quad() function, the *order* in which the vertices are listed *does* make a difference. For example, given the order in which the vertices are currently listed in the call to the quad() function in the second statement of our program,

```
size(150, 200);
quad(0, 0, 0, 60, 90, 75, 110, 35);
```

the order of our four vertices when the quadrilateral is drawn is considered to be

When the quadrilateral is drawn, the four line segments that are drawn connecting the four vertices are

From ❶ to ❷

From ❷ to ❸

From ❸ to ❹

From ❹ to ❶

Accordingly, if we change the order of the vertices in our function call,

```
size(150, 200);
quad(0, 0, 0, 60, 110, 35, 90, 75);
```

then, when we run our revised program, we now see somewhat of a "bow tie" shape:

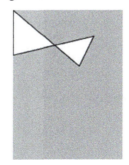

This is because the order of our vertices changed to

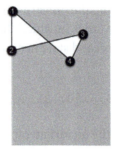

and the four line segments connecting the four vertices were drawn accordingly.

To avoid the bow tie effect, simply start the list of column and row arguments at a particular vertex and proceed in a clockwise (or counterclockwise) order until the column and row of all three of the other vertices have been listed.

Select **File > Save** to save this program. We will continue working with it in the next section.

Drawing Arcs: The `arc()` Function

Processing's **arc()** function allows us to draw *portions* of ellipses.

Recall that the basic form of the **ellipse()** function is

```
ellipse(column, row, width, height);
```

in which the four parameters are

column	The pixel column location of the ellipse's center
row	The pixel row location of the ellipse's center
width	The width of the ellipse in pixels
height	The height of the ellipse in pixels

Thus, the call to the **ellipse()** function in the following program,

```
size(150, 200);
ellipse(50, 60, 90, 75);
```

draws an ellipse with its center located at column 50 and row 60, a width of 90 pixels, and a height of 75 pixels:

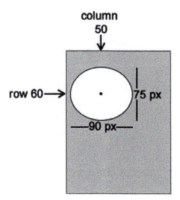

The first four parameters of the **arc()** function are the same as those of the ellipse() function. However, there are two additional parameters, a **starting angle** and an **ending angle**, that specify the portion of the ellipse that is to be drawn.

The angle measures for the starting and ending angle must be given in *radians* rather than degrees. Fortunately, Processing provides a **radians()** function for converting degrees to radians.

It is also important to note that the angle measures proceed in a *clockwise* manner, the reverse of the counterclockwise direction you may have learned in math class.*

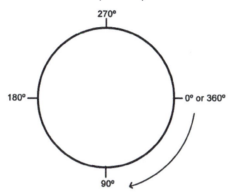

For example, let's change our call to the ellipse() function to the following call to the arc() function simply by changing the function name and by adding a starting angle and an ending angle, respectively:

```
size(150, 200);
arc(50, 60, 90, 75, radians(90), radians(180) );
```

Notice that we were able to specify our two angles using degrees, converting each angle measurement using the radians() function. As a result, there are now two calls to the radians() function *inside* our call to the arc() function. We are allowed to do this! Inserting a call to a function inside another call to a function like this is known as "nesting" function calls.

When we run our program, we now see the following result:

---

* This is because if we consider the columns and rows of pixels on the canvas to be like the *x* and *y* axes in geometry, then the *y* axis is "upside down" in comparison with geometry, because we move downward on the canvas rather than upward as the *y* values increase.

We see this result because the first four arguments,

```
arc(50, 60, 90, 75, radians(90), radians(180) );
```

define an ellipse,

and the last two arguments,

```
arc(50, 60, 90, 75, radians(90), radians(180) );
```

specify that we want only the portion of this ellipse that is between the 90° position and the 180° position to be drawn:

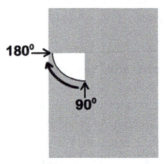

We can also describe the angles using *negative* degrees

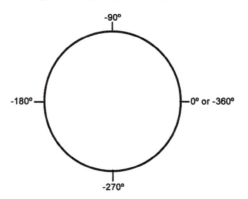

as long as the start angle and end angle describe a *clockwise* rotation. For example, the following call to the arc() function,

```
arc(50, 60, 90, 75, radians(-90), radians(90) );
```

will render the following arc:

We see this result because the starting angle was specified to be to be –90° and the ending angle was specified to be 90°. The use of a negative angle measure was allowed because the two angles still specified a clockwise rotation.

In summary, the basic form of the **arc()** function can be written as

```
arc(column, row, width, height, start, end);
```

in which the six parameters are

column	The pixel column location of the ellipse's center
row	The pixel row location of the ellipse's center
width	The width of the ellipse in pixels
height	The height of the ellipse in pixels
start	The angle at which the arc starts (given in radians)
end	The angle at which the arc ends (given in radians)

Select **File > Save** to resave the Drawing program.

Summary

In this section, we started by learning about a number of functions for drawing graphic elements in Processing.

```
size(width, height);
```
   Parameters
width	The desired pixel width of the canvas
height	The desired pixel height of the canvas

```
point(column, row);
```
   Parameters
column	The pixel column where the point is to be drawn
row	The pixel row where the point is to be drawn

```
line(column₁, row₁, column₂, row₂);
```
   Parameters
$column_1$	The pixel column location of the first endpoint
row_1	The pixel row location of the first endpoint
$column_2$	The pixel column location of the second endpoint
row_2	The pixel row location of the second endpoint

```
rect(column, row, width, height [, rounding] );
```
   Parameters
column	The pixel column location of the rectangle's upper-left corner
row	The pixel row location of the rectangle's upper-left corner
width	The width of the rectangle in pixels
height	The height of the rectangle in pixels
rounding	The amount of rounding of the corners (optional)

```
ellipse(column, row, width, height);
```
   Parameters
column	The pixel column location of the ellipse's center
row	The pixel row location of the ellipse's center
width	The width of the ellipse in pixels
height	The height of the ellipse in pixels

```
triangle(column₁, row₁, column₂, row₂, column₃, row₃);
```
   Parameters
$column_1$	The pixel column location of the first vertex
row_1	The pixel row location of the first vertex
$column_2$	The pixel column location of the second vertex
row_2	The pixel row location of the second vertex
$column_3$	The pixel column location of the third vertex
row_3	The pixel row location of the third vertex

```
quad(column₁, row₁, column₂, row₂, column₃, row₃, column₄, row₄);
```
   Parameters
$column_1$	The pixel column location of the first vertex
row_1	The pixel row location of the first vertex

$column_2$	The pixel column location of the second vertex
row_2	The pixel row location of the second vertex
$column_3$	The pixel column location of the third vertex
row_3	The pixel row location of the third vertex
$column_4$	The pixel column location of the fourth vertex
row_4	The pixel row location of the fourth vertex

```
arc(column, row, width, height, start, end);
```
Parameters

`column`	The pixel column location of the ellipse's center
`row`	The pixel row location of the ellipse's center
`width`	The width of the ellipse in pixels
`height`	The height of the ellipse in pixels
`start`	The angle at which the arc begins (given in radians)
`end`	The angle at which the arc ends (given in radians)

The Processing Reference

Additional information about all the above functions can be found in the **Reference** that is built into Processing and can be reached via the **Help** menu:

Some of the information is rather technical, so do not be worried if you don't understand all that is written in the Reference!

More about Graphical Elements

Let's learn more about working with graphical elements in Processing. Select **File > New** to start a new program. Select **File > Save** and save this new program as **MoreDrawing**.

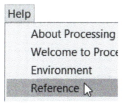

(You may **close** the *other* Processing window containing the Drawing program.)

Our MoreDrawing program doesn't yet contain any lines of code. Thus, when we run it, we see only a display window containing a blank gray canvas of the default size, 100 pixels by 100 pixels:

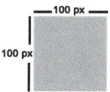

In the Processing window containing the `MoreDrawing` program, enter the following two lines of code:

```
rect(10, 15, 40, 70);
ellipse(65, 50, 55, 70);
```

When we run this program, we see that Processing simply draws a rectangle and an ellipse on a canvas of the default size (100 pixels by 100 pixels):

Select **File > Save** to resave this `MoreDrawing` program.

Throughout this section, we will repeatedly return to these two lines of code as our starting point.

Stacking Order

Notice that when we run our current program,

```
rect(10, 15, 40, 70);
ellipse(65, 50, 55, 70);
```

the ellipse is drawn *on top of* the rectangle.

We see this result because of what is sometimes known as the **stacking order**. In Processing, when a graphical element is drawn *after* some other graphical element(s), it is drawn *on top of* those graphical elements that were drawn previously.

In our current program, the call to the `rect()` function is the *first* statement in the program, and the call to the `ellipse()` function is the *second* statement of the program:

```
rect(10, 15, 40, 70);
ellipse(65, 50, 55, 70);
```

Therefore, when we run this program, the ellipse is drawn *after* the rectangle. Accordingly, the ellipse is drawn *on top of* the rectangle.

On the other hand, suppose we *reverse* the order of the statements in our program:

```
ellipse(65, 50, 55, 70);
rect(10, 15, 40, 70);
```

Now, when we run our program, the rectangle is drawn *after* the ellipse and, as a result, the rectangle is drawn *on top* of the ellipse:

Let's return our two statements to their original order:

```
rect(10, 15, 40, 70);
ellipse(65, 50, 55, 70);
```

When we run the program, the ellipse is once again drawn on top of the rectangle:

In some graphic design software, it is possible to change the stacking order after the graphical elements are drawn. However, in Processing, the only way to change the stacking order of graphical elements is by changing the order in which they are drawn.

This example illustrates a very important principle in computer programming: the performing of actions in a specified **sequence**. As is the case in a cooking recipe, the order in which the individual actions that make up a computer program are performed is significant. Changing the order of the actions may produce a different result.

Changing Line Thickness: The `strokeWeight()` Function

By default, the points, the line segments, and the edges around the shapes we have drawn have all been one pixel in thickness. This attribute of these graphical elements is known in Processing as the **stroke weight**.

Processing provides the **`strokeWeight()`** function that enables us to change the stroke weight used when points, line segments, or shape edges are subsequently drawn.

The general form of this function is

```
strokeWeight(pixels);
```

The only parameter of the `strokeWeight()` function is

    pixels   The desired stroke weight, specified in pixels

Consider once again our current program:

```
rect(10, 15, 40, 70);
ellipse(65, 50, 55, 70);
```

If we add the statement

```
rect(10, 15, 40, 70);
ellipse(65, 50, 55, 70);
strokeWeight(10);
```

then, when we run this program, we see the same result as before:

There is no change to the drawing that Processing produces because our call to the strokeWeight() currently take place *after* the calls to the rect() and ellipse() functions. In other words, we are changing the stroke weight too late, after the rectangle and ellipse have already been drawn. Only a point, line segment, or shape edge drawn *after* a call to the strokeWeight() function will have this new stroke weight.

Let's reposition the call to the strokeWeight() function so that it is *before* the call to the ellipse() function but *after* the call to the rect() function:

```
rect(10, 15, 40, 70);
strokeWeight(10);
ellipse(65, 50, 55, 70);
```

Now, when we run our program, only the ellipse is drawn with this new stroke weight:

On the other hand, if we position the call to the strokeWeight() function before *both* the call to the rect() function and the call to the ellipse() function,

```
strokeWeight(10);
rect(10, 15, 40, 70);
ellipse(65, 50, 55, 70);
```

then, when we run our program, *both* the rectangle and the ellipse are drawn with this new stroke weight:

There is an important general principle at work here: when we change a setting for the drawing of a graphical element, that change in the setting will *remain in effect* while we draw additional graphical elements. This is why *both* the rectangle and the ellipse were drawn with the new stroke weight: the current stroke weight remains in effect until it is reset using a new call to the strokeWeight() function.

If we would like the outline of the ellipse to have a different stroke weight, then we can simply insert a second call to the strokeWeight() function that comes *after* the call to the rect() function but *before* the call to the ellipse() function:

```
strokeWeight(10);
rect(10, 15, 40, 70);

strokeWeight(5);
ellipse(65, 50, 55, 70);
```

Now, when we run our program, the rectangle and the ellipse have different stroke weights:

Working with Color: RGB

So far, we have used only three pixel colors when drawing in Processing: black, white, and a shade of gray. However, Processing allows us to draw with colors as well.

Colors must be described as *numbers* in order for computers to be able to work with them. One of the most common ways that colors are numbered is according to the **RGB** color model, where all colors are described as mixed amounts red (R), green (G), and blue (B) light.

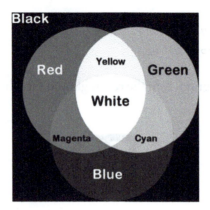

The three primary colors of light are red, green, and blue. When red light is mixed with green light, yellow light is produced. When red light is mixed with blue light, magenta light is produced. When green light is mixed with blue light, cyan light is produced. When red, green, and blue light are mixed in maximum amounts, white light is produced. When there is no red, no green, and no blue light at all, black is the result.

When the RGB model is used in computer technologies, the amounts of red, green, and blue light are typically described on a scale from 0 to 255. Thus, any color is described in RGB as a set of three numbers, each of which is from 0 to 255. For example, the RGB numbers for the basic colors in the above diagram are the following:

	Red	Green	Blue
Red	255	0	0
Green	0	255	0
Blue	0	0	255
Yellow	255	255	0
Magenta	255	0	255
Cyan	0	255	255
White	255	255	255
Black	0	0	0

Why 255? The amount of each of the red, green, and blue components is actually an eight-digit binary number, which is also known as 1 "byte." The largest binary number that can be stored in a byte is 11111111, which is the number we write in decimal as 255.

Processing also provides a built-in tool for determining the RGB numbers of a particular color. From the **Tools** menu, select **Color Selector**:

In the window that appears, we can click to select a certain color, and its RGB values will be displayed in the corresponding textboxes:

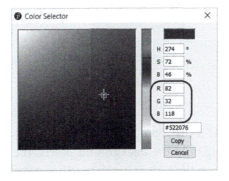

Resetting the Canvas: The `background()` Function

As we have seen, the default color of the canvas is a particular shade of gray. However, Processing provides the **`background()`** function, which resets all the pixels of the canvas to whatever RGB color we specify.

The basic form of a call to the `background()` function is

```
background(red, green, blue);
```

The parameters are

red	The amount of red (0 to 255)
green	The amount of green (0 to 255)
blue	The amount of blue (0 to 255)

Let's return to the starting version of our `MoreDrawing` program by removing the statements other than the following:

```
rect(10, 15, 40, 70);
ellipse(65, 50, 55, 70);
```

When we run this program, we once again see the following:

To change the background to green pixels, we might try the following:

```
rect(10, 15, 40, 70);
ellipse(65, 50, 55, 70);
background(0, 255, 0);
```

However, when we run this program, we see a canvas that consists entirely of green pixels:

Why is this the result? As it turns out, the background() function is somewhat misnamed. Instead of changing only the background pixels, the background() function actually resets *all* the pixels on the canvas to the specified color. For this reason, we need to set the background color of the canvas *before* we draw on the canvas:

```
background(0, 255, 0);
rect(10, 15, 40, 70);
ellipse(65, 50, 55, 70);
```

When we run this program, we now see the rectangle and ellipse drawn on top of a green canvas:

Changing the Fill Color: The fill() and noFill() Functions

By default, Processing fills the shapes that we draw—ellipse, rectangles, triangles, and quadrilaterals—with white pixels. This is known as the **fill color**.

Processing provides the **fill()** function, which allows us to change the fill color of any shape that we draw *after* we call this function.

The basic form of a call to the **fill()** function is

```
fill(red, green, blue);
```

The parameters are

red	The amount of red (0 to 255)
green	The amount of green (0 to 255)
blue	The amount of blue (0 to 255)

Let's once again return to the starting version of our `MoreDrawing` program by deleting all of the lines of code other than the following:

```
rect(10, 15, 40, 70);
ellipse(65, 50, 55, 70);
```

Recall, when we run this program, we see the following on the canvas:

Now, suppose that we would like to have the ellipse filled with yellow. To achieve this, we might try writing

```
rect(10, 15, 40, 70);
ellipse(65, 50, 55, 70);
fill(255, 255, 0);
```

However, when we run our program, we see the same result as before:

The ellipse that is drawn is not filled with yellow. This is because, given the order of the statements in our program, we set the fill color to yellow *after* the ellipse was already drawn. Thus, any *additional* shape that drawn *afterward* would be filled with yellow, but the fill color of any shape that has *already* been drawn will remain unchanged.

Once again, we see an important general principle at work here: when we wish to change a particular setting for drawing graphical elements, we must do this *before* we draw the graphical elements that we wish to be effected by this change.

In our current program, we need to move the current call to the `fill()` function so that it is *before* the call to the `ellipse()` function. If we position the call to the `fill()` function at the beginning of the program,

```
fill(255, 255, 0);
rect(10, 15, 40, 70);
ellipse(65, 50, 55, 70);
```

then both the rectangle and the ellipse are filled with yellow:

If we want to have only the ellipse filled with yellow, then we need to position the call to the `fill()` function *after* the call to the `rect()` function but *before* the call to the `ellipse()` function:

```
rect(10, 15, 40, 70);
fill(255, 255, 0);
ellipse(65, 50, 55, 70);
```

Now, when we run our program only the ellipse is filled with yellow, and the rectangle is filled with the default fill color, white:

It is important to remember the general principle that was previously mentioned:

> When we change a setting for the drawing of a graphical element, that change in the setting will *remain in effect* for any graphical elements that we draw *afterward*.

In the current program, any shapes that we draw after the ellipse will continue to be filled with yellow, unless we subsequently change the setting of the fill color using a call to the `fill()` function or to the `noFill()` function.

Processing also provides the **noFill()** function to *turn off* the filling of shapes. This function does not have any parameters. Thus, the general form of a call to this function is

```
noFill();
```

For example, let's add a call to the `noFill()` function at the beginning of our program:

```
noFill();
rect(10, 15, 40, 70);
fill(255, 255, 0);
ellipse(65, 50, 55, 70);
```

Notice that, as a result, the rectangle is now unfilled and, instead, the default color of the canvas now shows through on the inside of the rectangle:

In contrast, notice that the ellipse remains filled with yellow. This is because there is a call to the `fill()` function before we draw the ellipse:

```
noFill();
rect(10, 15, 40, 70);

fill(255, 255, 0);
ellipse(65, 50, 55, 70);
```

A call to the `fill()` function not only sets the fill color but also causes the filling of shapes to *resume*.

Changing the Stroke Color: The `stroke()` and `noStroke()` Functions

Notice that all of the points, line segments, and the edges of the shapes we have drawn in Processing have been black. This is because black is the default **stroke color**.

Processing provides the **`stroke()`** function that allows us to change the stroke color that will be used for any points, line segments, or shape edges that we draw *after calling this function.*

The basic form of a call to the **`stroke()`** function is

```
stroke(red, green, blue);
```

The parameters are

red	The amount of red (0 to 255)
green	The amount of green (0 to 255)
blue	The amount of blue (0 to 255)

Let's once again revise our `MoreDrawing` program to consisting of just the following two lines of code:

```
rect(10, 15, 40, 70);
ellipse(65, 50, 55, 70);
```

We'll now add a call to the `strokeWeight()` function at the beginning of our program:

```
strokeWeight(5);
rect(10, 15, 40, 70);
ellipse(65, 50, 55, 70);
```

As a result, when we run this program, the edges around our shapes are much more visible:

Next, let's place a call to the `stroke()` function at the beginning of our program that sets the stroke color to red:

```
stroke(255, 0, 0);
strokeWeight(5);
rect(10, 15, 40, 70);
ellipse(65, 50, 55, 70);
```

Thus, before we draw our shapes, we are now making changes to *two* stroke settings: the stroke **color** and the stroke **weight**. As a result, when we run our program, we see that each of our shapes is now drawn with a red stroke around it that has a thickness of 5 pixels:

Remember, the changing of a drawing property stays in effect for any shapes that are drawn *afterward*. Thus, in the current program, any stroking—that is, the drawing of points, line segments, or shape edges—that takes place *after* this call to the `stroke()` function will also be red in color and 5 pixels in thickness. This is why *both* the rectangle and the ellipse have been drawn with the specified stroke weight and stroke color.

Processing also provides the **noStroke()** function to *turn off* the drawing of edges around shapes. This function does not have any parameters. Thus, the general form of a call to this function is

```
noStroke();
```

To illustrate, let's change the call to the `stroke()` function at the beginning of our program to a call to the `noStroke()` function:

```
noStroke();
strokeWeight(5);
rect(10, 15, 40, 70);
ellipse(65, 50, 55, 70);
```

As a result, when we run our program, neither the rectangle nor the ellipse has an edge drawn around it:

Once the `noStroke()` function is called, no stroking will take place. There will be no drawing of points, line segments, or shape edges until there is a call to the `stroke()` function. For example, if we make the following change to our program,

```
noStroke();
strokeWeight(5);
rect(10, 15, 40, 70);
stroke(255, 0, 0);
ellipse(65, 50, 55, 70);
```

then the rectangle still does not have a stroke around it, but there is once again a red stroke around the ellipse because it is drawn *after* the call to the `stroke()` function:

Inline Comments

Processing gives us the option of putting **comments** in our code. One common use of comments is to clarify the purpose of one or more lines of our code, not only for others who might read our code but also for ourselves as we work on a program.

One form of commenting in Processing with is **inline** comments. If we type two forward slashes (//), Processing will *ignore* anything we write from the point where the slashes are inserted up to the end of that line of code.

For example, to clarify that the call to the `stroke()` function in our current program will turn on red stroking, we can add the following comment:

```
noStroke();
strokeWeight(5);
rect(10, 15, 40, 70);
stroke(255, 0, 0); // red stroke
ellipse(65, 50, 55, 70);
```

Notice that this does not cause an error, because Processing ignores everything we have written from the two slashes to the end of that line of code.

Grayscale

Let's once again edit our `MoreDrawing` program so that it consists of just the following two statements:

```
rect(10, 15, 40, 70);
ellipse(65, 50, 55, 70);
```

When we run this program, we once again see the following drawn on the canvas:

Our canvas color is currently the default shade of gray. As it turns out, the RGB values for this particular shade of gray are **204, 204, 204**.

Suppose that we wish to set our background to a lighter shade of gray instead. For example, we could do so by inserting the following call to the background()function at the beginning of our program:

```
background(240, 240, 240);
rect(10, 15, 40, 70);
ellipse(65, 50, 55, 70);
```

When we run this program, we now see a lighter shade of gray for the canvas:

Notice that, once again, all three of the RGB values for this shade of gray are the same: 240, 240, 240.

Actually, like the RGB numbers for black—0, 0, 0—and the RGB numbers for white—255, 255, 255—*all three* of the RGB numbers for a shade of gray are always the *same*. Whenever red, green, and blue light are mixed in an equal amount that is greater than 0 and less than 255, some shade of gray light is produced.

Another way of visualizing this option is in terms of a **grayscale**, running from black to white through many shades of gray. In Processing, any color on this grayscale is described by a set of three identical RGB numbers, from 0, 0, 0 (black) to 255, 255, 255 (white):

0,0,0   255,255,255

Because all three of the RGB numbers for any color on this grayscale are always the same, Processing also permits us to refer to black, white, or a particular shade of gray using just *one* of the three identical RGB numbers. Thus, in our current example, instead of our current statement

```
background(240, 240, 240);
```

we actually have the option of simply writing

```
background(240);
```

Thus, we can also think of the color numbers of this grayscale as a range of single numbers from 0 to 255.

However, for consistency, in this book we will always use a set of three RGB numbers to describe any color, even when that color is black, white, or a shade of gray.

Transparency

Let's once again return our `MoreDrawing` program to the starting point that consists of just the following two lines of code:

```
rect(10, 15, 40, 70);
ellipse(65, 50, 55, 70);
```

When we run this program, we once again see the following drawn on the canvas:

Let's again set the fill color of the ellipse to yellow using the following call to the `fill()` function:

```
rect(10, 15, 40, 70);
fill(255, 255, 0);
ellipse(65, 50, 55, 70);
```

When we run this program, the fill color of the ellipse is indeed set to yellow:

As we have seen, an RGB color description consists of three numbers, each of which is from 0 to 255. This describes a color that is totally **opaque**, not at all transparent. Thus, in our current example, the portion of the rectangle that the ellipse overlaps is completely invisible.

However, Processing also allows us the option of adding a *fourth* number to describe the **transparency** of the selected color. This fourth number is also in the range from 0 to 255, where 0 is a color that is totally transparent and 255 is a color that is totally opaque (the default). This number is sometimes known as the color's **opacity**, a term that refers to how *opaque* the color is. It is also known as the **alpha** value, and accordingly, this use of four numbers in this manner is sometimes known as the **RGBA** color model.

In our current example, the call to the `fill()` function that sets the fill color of our ellipse is

```
fill(255, 255, 0);
```

If we would prefer to have our ellipse filled instead with a *semitransparent* yellow, we simply add a *fourth* argument to the call to the `fill()` function. This fourth argument can be any value between 0 (totally transparent) and 255 (totally opaque). For example, we can write

```
rect(10, 15, 40, 70);
fill(255, 255, 0, 150);
ellipse(65, 50, 55, 70);
```

Now, when we run our program, the ellipse is filled with a semitransparent yellow, and as a result, the portion of the rectangle that the ellipse overlaps is now partially visible:

Such transparency can also be used with the `stroke()` function to create a point, line segment, or shape edge that is semitransparent.

However, using the four-number RGBA model currently *has no effect* in Processing if it is used with the `background()` function.

Summary

In this section, we learned about a number of additional functions that can be used for drawing graphic elements in Processing.

`strokeWeight(pixels);`
    Parameters
        pixels   The desired stroke weight, specified in pixels

`background(red, green, blue);`
    Parameters
        red   The amount of red
        green   The amount of green
        blue   The amount of blue

`fill(red, green, blue);`
    Parameters
        red   The amount of red
        green   The amount of green
        blue   The amount of blue

```
noFill();
```
Parameters: None

```
stroke(red, green, blue);
```
Parameters

red	The amount of red
green	The amount of green
blue	The amount of blue

```
noStroke();
```
Parameters: None

Additional information about all these functions can also be found in the **Reference** that is built into Processing and can be reached via the **Help** menu:

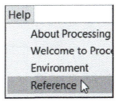

Processing has much more capability for drawing than was covered in this chapter. However, you are now well prepared to understand the graphical examples in the chapters ahead. You are also now in a great position to start having fun experimenting with drawing in Processing on your own.

Exercises

1) Write code that will draw the following on a 150-pixel by 200-pixel canvas:

2) Write code that will draw the following on a 150-pixel by 200-pixel canvas:

3) Write code that will draw the following on a 150-pixel by 200-pixel canvas:

4) Write code that will draw the following on a 150-pixel by 200-pixel canvas:

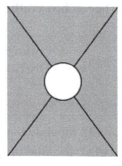

5) Write code that will draw the following on a 150-pixel by 200-pixel canvas:

6) Write code that will draw the following on a 150-pixel by 200-pixel canvas:

7) Write code that will draw the following on a 150-pixel by 200-pixel canvas:

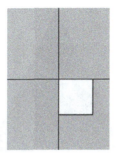

8) Using the ellipse() or arc() function, write code that will draw the following on a 200-pixel by 200-pixel canvas:

9) Using the ellipse() or arc() function, write code that will draw the following on a 200-pixel by 200-pixel canvas:

10) Using the ellipse() or arc() function, write code that will draw the following on a 200-pixel by 200-pixel canvas:

11) Write code that will draw the following on a 200-pixel by 200-pixel canvas:

12) Use the quad() function to draw a diamond shape like the following on a 200-pixel by 200-pixel canvas:

13) Write code that will draw the following on a 200-pixel by 200-pixel canvas:

14) Use the arc() function to draw the following on a 200-pixel by 200-pixel canvas:

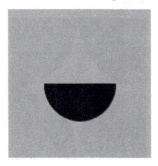

15) Write code that will draw the following on a 200-pixel by 200-pixel canvas:

16) Write code that will draw the following on a 200-pixel by 200-pixel canvas:

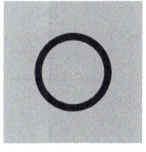

17) Write code that will draw the following on a 200-pixel by 200-pixel canvas:

18) Draw a simple house using Processing code. (Tip: Draw on graph paper first.)

Example

19) Draw a toy car using Processing code.

Example

20) Draw a snowman using Processing code.

Example

21) Draw an animal using Processing code.

Example

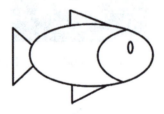

22) Create your initials using Processing code.

23) Pick a national flag with a simple design and try to recreate it with Processing code.

24) Try to recreate the Olympic flag with Processing code.

25) Create a drawing in the style of the painter Piet Mondrian using Processing code.

CHAPTER 2

Types, Expressions, and Variables

This chapter introduces basic concepts that are foundational to the chapters that follow.

In this chapter, we will learn how to work with various *types* of information in Processing. We will start by learning about some of the *types* of numbers that we can use, how we can do arithmetic with these types of numbers, and how we can store these types of numbers in *variables*. We will also learn how to work with nonnumerical items of information, such as single text characters and strings of text characters.

The graphics in this chapter are kept quite simple to make it easier to focus on the key concepts that are being introduced.

Values

In computing, a single item of information that is in its *simplest form* is known as a **value**.

As we shall see, we are able to work with different *types* of values in Processing. These include both *numeric* values such as an integer and *nonnumeric* values such as a letter of the alphabet or an entire word.

We will start with numeric values and look at nonnumeric values later in this chapter.

Numeric Values

Numeric values are specific numbers that are in their simplest form.

Examples

2 is a numeric value.

2.5 is a numeric value.

2 + 5 is not a numeric value, because it can be further simplified to 7.

7 is a numeric value.

Processing allows us to work with several **types** of numeric values. The two most commonly used numeric types in Processing are the **int** type and the **float** type.

Integers: The int Type

In Processing, whenever we use an integer (negative, zero, or positive), we are using what is known as an **int** value.

An **int** value can be any integer in the range from −2,147,483,648 to 2,147,483,647 but cannot have a decimal point.* Examples of **int** values:

```
1
10
123456789
-99
100
```

Numbers with Decimal Points: The float Type

Processing also allows us to work with values that are of the **float** type.

A **float** value is identifiable by its decimal point (also known as a "floating point"). A **float** value can be a much larger positive or negative number than is possible for an **int** value.† Scientific notation is also permitted for **float** values. For example, the **float** value 1000000.0 can also be written as 1E6 or as 1.0E6 in scientific notation.

Examples of **float** values:

```
10.0
2.5
0.3
-23.25
5.125
1.0E10
```

Arithmetic with int Values and float Values

Now that we have a basic understanding of the **int** type and the **float** type, let's look at how we can do *arithmetic* with these two types of numeric values.

The **operators** that Processing uses for arithmetic operations are the same as those used in many other programming languages, in spreadsheet software, and on some handheld calculators:

Multiplication   *

Division   /

---

* In other words, an int value can be in the range from -2^{31} to $2^{31} - 1$.

† A float value can be in the range from −340,282,347,000,000,000,000,000,000,000,000,000,000.0 to 340,282,347,000,000,000,000,000,000,000,000,000,000.0.

Addition   +
Subtraction   –

The numbers that we use with these operators are known as **operands**.
Thus, when we add 3 and 2,

    3 + 2

the *operation* is addition, the *operator* is the + sign, and the *operands* are 3 and 2.

int Arithmetic

Open Processing and start a new sketch by selecting **File > New**.
Save this new sketch as **Arithmetic**.

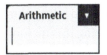

Processing is similar to many other programming languages in the way it performs arithmetic with integers—that is, with int values. The important principle to bear in mind regarding integer arithmetic in Processing is this:

> **If—and only if—both of the operands in an arithmetic operation are of int type, then the result will also be an int value.**

This is a fairly straightforward principle for multiplication, addition, and subtraction with integers. Examples:

    5 * 7 result: 35 (not 35.0)
    4 + 6 result: 10 (not 10.0)
    8 - 3 result: 5   (not 5.0)

Processing's **println()** function displays a line of information in the console and then advances to the next line in the console. (We will learn more about this function later in this chapter.) We can nicely test the results of operations with calls to the println() function. For, example, enter the following statement into the Text Editor area of the Processing window:

    println(5 * 7);

When we run this program, Processing performs the multiplication operation 5 * 7 and displays the result in the console:

    35

Likewise, if we enter the following statements,

```
println(4 + 6);
println(8 - 3);
```

and run this program, Processing performs the addition and subtraction operations, and we see the following results in the console output:

```
10
5
```

Integer Division

As we have learned, in Processing (as in most programming languages), any arithmetic operation in which *both* of the operands are int values always produces an int value. This includes division. However, we need to be careful when we perform *division* with int values.

To illustrate how Processing performs division with int values, it may be helpful to think of the kind of method you probably used when you first learned to do division in school, entirely with integers. For example, using this method to perform 16 ÷ 5, you begin with

$$5\overline{)16}$$

Next, you determine the largest number that can be multiplied by 5 without going over 16. This turns out to be 3. Thus, you multiply 5 by 3 to get 15. You then subtract 15 from 16 and discover that you have a remainder of 1.

$$
\begin{array}{r}
3 \\
5\overline{)16} \\
-15 \\
\hline
1
\end{array}
$$

Thus, the answer to this division operation is "3 remainder 1." You can write this answer in such a form as

Integer division in Processing operates in a similar way. However, there is a key difference: in Processing, the remainder is of an integer division operation is essentially *discarded*. For example, the operation

```
16 / 5
```

produces 3 as its result. We can verify this by typing the following program into the Text Editor area of the Processing window:

```
println(16 / 5);
```

When we run this one-line program, we see the following output in the console:

```
3
```

On a calculator, we would get 3.2 as the result of this division operation. However, in Processing, whenever an int value does *not* evenly divide another int value, an int value is the result and any fractional remainder is discarded. It's as if the fractional portion of the quotient that we would normally get is simply chopped off. Thus, the result of 16 / 5 in Processing is 3 (*not* 3.2).

Here are some other examples:

```
4 / 2  result: 2
5 / 2  result: 2  (not 2.5)
1 / 2  result: 0  (not 0.5)
```

We can test these integer division operations by placing them in calls to the println() function:

```
println(4 / 2);
println(5 / 2);
println(1 / 2);
```

When we run the program containing these statements, we see the following results of these integer division operations output to the console:

```
2
2
0
```

It is very important to remember that integer division works this way in Processing. Dividing one int value by another int value *always* produces a quotient that is an int value.

Calculating the Remainder with the Modulo Operator: %

Thinking of how you did integer division when you were younger is also helpful in another way. The previous example we considered,

helps to illustrate why the integer division operation 16 / 5 produces a quotient of 3 in Processing (and not 3.2), but it also reminds us how we worked with **remainders** when we first learned to do division entirely with integers.

Like many programming languages, Processing includes the **modulo** operator % for calculating the **remainder** that results from an integer division operation.

We now know that, in Processing, the operation 16 / 5 produces a quotient of 3 and essentially discards a remainder of 1. However, this remainder after performing integer division is exactly what Processing's modulo operator (%) gives us. To illustrate, consistent with our current example, the modulo operation

```
16 % 5
```

does indeed produce the following result in Processing:

```
1
```

We can verify this result by typing the following program into the Text Editor area of the Processing window:

```
println(16 % 5);
```

When we run this one-line program, we do indeed see the following remainder output to the console:

```
1
```

Here are some other examples of the modulus (remainder) operation:

```
4 % 2   result: 0     (4 ÷ 2 = 2 with a remainder of 0)
5 % 2   result: 1     (5 ÷ 2 = 2 with a remainder of 1)
8 % 5   result: 3     (8 ÷ 5 = 1 with a remainder of 3)
```

If we test these operations with calls to the `println()` function, as in the following program,

```
println(4 % 2);
println(5 % 2);
println(8 % 5);
```

then we see the following remainders output to the console:

```
0
1
3
```

float Arithmetic

Processing is also similar to many other programming languages in the way it performs arithmetic involving `float` values (numbers with decimal points). The important principle to bear in mind when performing arithmetic with `float` values in Processing is this:

> If *one or both* of the *operands* is a `float` value, then the *result* will also be a `float` value.

In other words, if even *one* of the operands has decimal places, then the *result* will also have decimal places.

Examples

```
5.0 + 2.0    result: 7.0
5.0 + 2      result: 7.0
5 + 2.0      result: 7.0

5.0 * 2.0    result: 10.0
5.0 * 2      result: 10.0
5 * 2.0      result: 10.0
```

```
5.0 - 2.0    result: 3.0
5.0 - 2      result: 3.0
5 - 2.0      result: 3.0

5.0 / 2.0    result: 2.5
5.0 / 2      result: 2.5
5 / 2.0      result: 2.5
```

Here, too, we can verify such results with calls to the `println()` function, as in the following program:

```
println(5.0 / 2.0);
println(5.0 / 2);
println(5 / 2.0);
```

When we run this program, we see the following quotients output to the console:

```
2.5
2.5
2.5
```

float *Fractions*

Arithmetic operations with `float` values *usually* produces results that are similar to what is produced if we perform the same arithmetic on a calculator. However, this is not always the case. When using `float` values, there is a very important fact that we need to remember:

Computers sometimes produce inaccurate results when fractions are involved.

As evidence of computer inaccuracy in working with fractions, note that the following statement,

```
println(1.0 / 3.0);
```

produces the following output to the console:

```
0.33333334
```

The 4 in the last digit is obviously incorrect. Similarly, the following statement,

```
println(0.6 + 0.1);
```

produces the following incorrect output to the console instead of 0.7:

```
0.70000005
```

When programming in Processing, we need to remember to be careful regarding `float` values, especially ones that have more than a few digits to the *right* of the decimal point.

In this book, we will use integers (`int` values) as often as possible.

The Trouble with Fractions on Computers

Are you curious about why computers have an inherent inaccuracy when it comes to fractions? This section explains a bit more about this problem. Don't worry if you don't understand everything that is covered.

Consider how, given enough digits, any whole number can be stored exactly on a calculator. However, many fractions such as ⅓ cannot be represented exactly in decimal digits. For example, on a calculator, 1 ÷ 3 produces something like

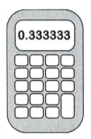

This inaccuracy results because of the limitations of the decimal (also known as "base-10") number system, which is based upon powers of 10.

Given enough digits to the *left* of the decimal point, any *whole number* can be represented *exactly* in decimal as some combination of quantities of powers of 10: ones (10^0), tens (10^1), hundreds (10^2), thousands (10^3), and so on.

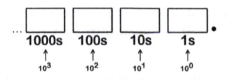

Because 10 is the base, any digit can count from 0 up to 9 before carrying to the next digit to the left. For example, fourteen is written as 14, because it is a combination of a 10 and four 1s.

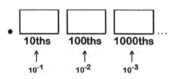

However, to the *right* of the decimal point, *fractions* must be represented in decimal by combining quantities of *negative* powers of 10. In other words, we must be able to combine quantities of tenths (10^{-1}), hundredths (10^{-2}), thousandths (10^{-3}), and so on, in such a way that this combination equals a given fraction.

For example, the following quantities of these decimal fractions combine to equal ¾:

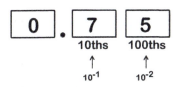

However, there is simply no way to combine quantities of these decimal fractions in a way that is ever exactly equal to one-third:

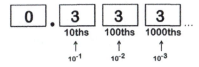

Rather, the result is a repeating decimal digit, which is sometimes written as

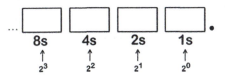

Internally, computers do not use the decimal (base-10) number system to represent numbers. Rather, computers use **binary** (also known as "base-2") representation, where only powers of 2 are used to represent the digits of a number. Because 2 is the base, any binary digit can count only from 0 up to 1 before carrying to the next binary digit to the left. Thus, a binary digit can only ever be a 0 or a 1. A binary digit is also known as a **bit**.

Why are binary numbers are used to build computers? It is easier to represent a binary digit on a computer, because a device with only two possible states is all that is needed. Consider an on–off switch: the "on" position can represent a 1 in a binary digit, and the "off" position can represent a 0 in a binary digit.* Thus, computers are built using a variety of two-state devices such as transistors (which are used in computers as very fast on–off switches).

Given enough digits to the *left* of the binary point, any *whole number* can be exactly represented in binary as some combination of positive powers of 2: ones (2^0), twos (2^1), fours (2^2), eights (2^3), and so on.

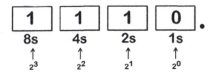

For example, the number 14 would be written in binary as 1110, because it is a combination of an 8, a 4, and a 2:

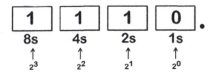

---

* Representing a decimal (base-10) digit requires a more complicated device with 10 possible states (e.g., a gear with 10 positions) so that it can represent 0 through 9.

With eight binary digits, we can represent 256 different values, because we can count from 00000000 up to 11111111, the number we write in decimal as 255:

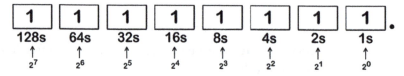

This is why the RGB values we use are from 0 to 255: because 128 + 64 + 32 + 16 + 8 + 4 + 2 + 1 = 255. Behind the scenes, Processing actually uses an eight-digit binary number (in other words, 1 byte) to store each of the three RGB components—eight binary digits for red, eight binary digits for green, and eight binary digits for blue.* An eight-digit binary number is also known as a **byte**.

However, what about binary *fractions*? Using digits to the *right* of the binary point, binary fractions must be represented as a combination of quantities of *negative* powers of 2. In other words, we must be able to combine quantities of halves (2^{-1}), fourths (2^{-2}), eighths (2^{-3}), sixteenths (2^{-4}), thirty-seconds (2^{-5}), and so on, in such a way that equals a given fraction.

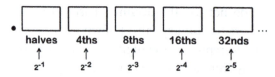

This means that, the only fractions that can be represented *exactly* as binary digits are $1/2$, $1/4$, $1/8$, $1/16$, and so on, along with any fractions that are *combinations* of these. For example, ¾ is written in binary digits as 0.11 because it is equal to ½ + ¼:

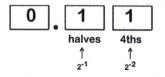

However, the overwhelming majority of fractions *cannot* be represented as combinations of these binary fractions. Thus, the problem of inaccuracy in representing *binary* fractions is much worse than the problem of representing decimal fractions accurately. For example, such a basic fraction as $1/10$ cannot be stored exactly in binary:

0.00011001100110011001100110011001100110011001100110011001100110011001100110...

Rather, the result is a repeating binary fraction:

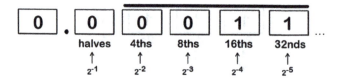

---

* Because a binary digit is also known as a bit, this RGB color scheme is also known as 24-bit color. This color scheme can represent 16,777,216 different combinations of red, green, and blue (256 × 256 × 256). Color transparency (RGBA) requires an additional eight binary digits (bits). Thus, 32 binary digits (bits) are required to describe each pixel of the Processing canvas.

It's okay if you don't fully understand how the binary representation of fractions works. However, do remember this: *very few fractions* can be stored exactly as binary digits and, as a result, computers have an inherent inaccuracy when fractions are involved!

Evaluating Expressions

When we combine two or more operands and one or more arithmetic operators, we produce what is known as an arithmetic **expression**.

Examples

```
5 * 2
5 * 2 * 3
10.1 + 4.5
8 - 5 - 2
```

Working out the result of an expression is also known as **evaluating** that expression, because it produces a single *value* as a result.

Examples

```
5 * 2          evaluates to 10
5 * 2 * 3      evaluates to 30
10.1 + 4.5     evaluates to 14.6
8 - 5 - 2      evaluates to 1
```

As we have seen, we can use the `println()` function to test the evaluation of expressions. For example, if we enter the following

```
println(8 - 5 - 2);
```

into the Text Editor area of the Processing window and then run this one-line program, Processing will evaluate the expression and display the resulting value. Thus, in this particular case, we see the following resulting value output to the console:

```
1
```

This output confirms that the expression 8 − 5 − 2 does indeed evaluate to 1 in Processing.

Order of Operations

In math class, you may have learned a "PEMDAS" acronym (e.g., "Please Excuse My Dear Aunt Sally") to help you remember the order in which the operations in an arithmetic expression are performed: *P*arentheses, *E*xponents, *D*ivision, *M*ultiplication, *A*ddition, *S*ubtraction. These rules for the order in which the arithmetic operations are performed are also known as **operator precedence**.

The order of operations (operator precedence) in Processing works in a similar way. However, one key difference is that, like many programming languages, Processing does not have a symbol that can be used for indicating an exponent.* Thus, when two or more operators are combined in a single expression, Processing performs these operations in the following order:

1. Parentheses
2. Multiplication or division
3. Addition or subtraction

Examples

```
3 * 2 + 5     evaluates to 11
3 + 2 * 5     evaluates to 13
3 + 2) * 5    evaluates to 25
```

The results of each of these expressions can be verified with calls to the `println()` function, as in the following program:

```
println(3 * 2 + 5);
println(3 + 2 * 5);
println((3 + 2) * 5);
```

When we run this program, we see the following resulting values output to the console:

```
11
13
25
```

In Processing, whenever there are two operators within an expression that have the same level of precedence, these operations are performed from *left to right*. This left-to-right order of operations can make a difference when integer division is involved.

For example, consider the expression

```
100 * 3 / 4
```

If we test this expression using the `println()` function

```
println(100 * 3 / 4);
```

we see from the console output that this expression evaluates to

```
75
```

---

* As we'll learn later, Processing provides the `pow()` function for the purpose of raising a number to a power. For example, 3^2 would be coded as `pow(3, 2)`.

Because multiplication and division have the same level of precedence, these two operations in this expression are performed from left to right. Thus, the multiplication is performed before the division. On the other hand, consider the expression

```
3 / 4 * 100
```

If we test this expression using the `println()` function

```
println(3 / 4 * 100);
```

we see from the console output that this expression evaluates to

```
0
```

In this expression, the division is performed before the multiplication. Because 3 / 4 is *integer* division, the resulting quotient is zero. Thus, when the multiplication operation is performed next, the resulting product is also zero.

Using Parentheses

Consider using *parentheses* whenever a single expression contains two or more operators. Parentheses not only remove the need to rely on the default rules for the order of operations but also improve the *readability* of the expression.

For example, in the following expression,

```
3 * 2 + 5
```

parentheses are not required for the multiplication operation to be performed before the addition operation, because multiplication has higher precedence than addition. However, if parentheses are added anyway, the result is an expression that is less ambiguous and easier to read:

```
(3 * 2) + 5
```

Variables

So far, we have been using specific values in our program statements.
Examples:

10 is an example of an `int` value.
10.5 is an example of a `float` value.

In addition to using specific values, Processing allows us to create and use *variables*. A **variable** in computer programming has some similarity to the way that a variable is used in algebra, where an expression such as

$$y = x + 10$$

uses the variable *x* as a kind of placeholder for a number that will be filled in later. It is as if there is an empty box named "*x*" into which a number will be inserted later:

$$y = \boxed{}^{x} + 10$$

For example, if we supply 9 as the value of *x*,

$$y = \boxed{9}^{x} + 10$$

then our expression produces 19 as the resulting value that is assigned to *y*.

A variable used in computer programming is also analogous to a cell in a spreadsheet. A computer's memory is actually somewhat like a spreadsheet, in that it is comprised of many individual storage locations, each of which has a unique address. For example, in the following spreadsheet, the cell with the address A1 currently contains the value 25:

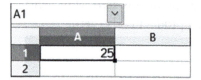

In this spreadsheet, we could create a formula that makes use of the value stored in this cell. For example, the formula

```
= A1 * 10
```

would produce 250 as its result.

However, remembering cell addresses can be tedious. Fortunately, most spreadsheet programs also allow us to give a *name* to a cell. For example, in the case of the following spreadsheet, the cell at location A1 has been given the name "total":

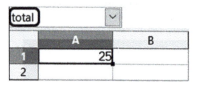

Thus, when formulas are created in this spreadsheet, the cell in address location A1 can now be referenced using the name "total" instead. For example, the formula

```
= total * 10
```

would still produce 250 as its result.

Similarly, in computer programming, a variable is a *storage location*. Behind the scenes, each variable has a unique address (a unique binary number) that describes its location within the computer's memory. Fortunately, each variable is also given a *name*. Thus, it is not necessary for us to know the actual address of this memory location. Rather, as we'll see, we can simply refer to this variable using its name whenever we need to *store* a value in that variable or *retrieve* a copy of the value currently stored in that variable.

Save and close the `Arithmetic` program.

Predefined Variables: `width` and `height`

In Processing, select **File > New** to start a new program. Save this new program as **WidthHeight** .

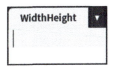

There are actually several useful variables that are *already* built into Processing. Let's look at two of these variables. Processing's `width` and `height` variables keep track of the canvas size. The **width** variable stores the number of *columns* of pixels that make up the current canvas. Similarly, the **height** variable stores the number of *rows* of pixels that make up the current canvas.

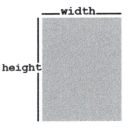

Each of these two variables, `width` and `height`, is a named storage location that can store a single `int` value. As we will learn, such a variable can only store one specific *type* of value. We can't actually see these two named storage locations, but we can visualize them as

As we know, the default size of the canvas for a Processing sketch is comprised of 100 columns of pixels and 100 rows of pixels.

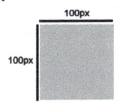

Thus, when we run a Processing program, the default number of canvas columns of pixels (100) is initially stored in the `width` variable, and the default number of canvas rows of pixels (100) is initially stored in the `height` variable.

We can verify this using the `println()` function. Let's enter the following program into the Text Editor area of the Processing window:

```
println(width);
println(height);
```

When we run this two-line program, it still produces a canvas that is the default 100 by 100 pixels in size:

However, we also now see the following output in the console showing the values currently stored in the `width` and `height` variables:

```
100
100
```

Next, let's insert the following call to the `size()` function into our program:

```
size(150, 200);
println(width);
println(height);
```

Now, when we run this program, the value 150 is automatically stored in Processing's `width` variable, and the value 200 is automatically stored in Processing's `height` variable.

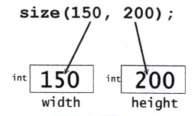

We see a canvas rendered that is the new size we specified:

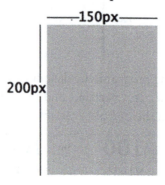

We also see the following output to the console, showing the new values that are stored in the `width` and `height` variables:

```
150
200
```

The `width` and `height` variables are useful in quite a variety of ways. For example, suppose that we want to draw a circle with a diameter of 40 pixels at the center of the current canvas. We can use Processing's `ellipse()` function for this. As we learned in Chapter 1, the `ellipse()` function requires us to specify four items of information (arguments). The first two arguments specify the pixel column and pixel row, respectively, of the center of the ellipse to be drawn. Thus, to place our circle in the center of the current canvas, we can simply calculate half of the canvas width (150 ÷ 2 = 75) and half of the canvas height (200 ÷ 2 = 100), and use these specific values, 75 and 100, as the pixel location for the center of our circle:

```
ellipse(75, 100,  );
```

The third and fourth arguments we supply to the ellipse function specify the dimensions of the ellipse we desire. Thus, for a circle with a diameter of 40 pixels, we will add

```
ellipse(75, 100, 40, 40);
```

Our program is now

```
size(150, 200);
println(width);
println(height);
ellipse(75, 100, 40, 40);
```

When we run this program, we see that a circle with a diameter of 40 pixels is indeed drawn in the center of the canvas:

The console output of the values of the `width` and `height` variables is still

```
150
200
```

Now, suppose we enlarge the size of our canvas:

```
size(220, 250);
println(width);
println(height);
ellipse(75, 100, 40, 40);
```

When we run our program, the console output of the values of the width and height variables is now

```
220
250
```

However, we find that the circle is no longer at the center of the canvas:

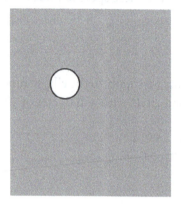

The circle is no longer centered because the specific values we calculated and used for the center of the circle in our ellipse() function, 75 and 100, are no longer equal to half of the width of the canvas and half of the height of the canvas. Rather, the horizontal center of the current canvas is now at column 110 (because 220 ÷ 2 = 110), and the vertical center of the current canvas is now at row 125 (because 250 ÷ 2 = 125).

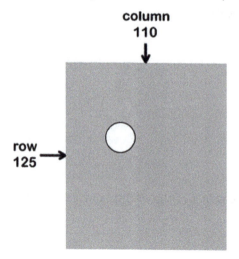

A more versatile technique for positioning our circle in the center of the canvas is for us to use the width and height variables in place of these specific values. We can describe the *horizontal* center of the canvas with the expression

```
width / 2
```

And we can describe the *vertical* center of the canvas with the expression

```
height / 2
```

If we use these two expressions to specify the pixel column and pixel row for the center of our circle, then our program becomes

```
size(220, 250);
println(width);
println(height);
ellipse(width / 2, height / 2, 40, 40);
```

When we run this program, the console output of the width and height variables is still

```
220
250
```

However, we see that the circle is once again drawn at the center of the canvas:

Furthermore, look what happens when we change the size of the canvas back to its previous size:

```
size(150, 200);
println(width);
println(height);
ellipse(width / 2, height / 2, 40, 40);
```

When we run this program again, we now see the following output to the console, showing the values stored in the width and height variables:

```
150
200
```

Also, even though the canvas size has changed, the circle remains at the center of the canvas:

Thus, by using expressions involving the `width` and `height` *variables* to position our circle at the center of the canvas instead of using specific values, we ensure that even if there is a change in the width and/or height of the canvas, the circle will remain at the center of the canvas. As we saw, this is *not* the case if we use *specific values* to set the pixel column and pixel row for the center of our circle. As we have already seen, variables allow for *variation* in our programs!

Benefits of Using Variables

As our example making use of the `width` and `height` variables illustrates, variables provide a way to describe the objectives of our programs in more *general* terms. By using variables, even if certain factors change, we can still obtain the results we desire. In other words, our program is more **flexible**, less likely to fail when certain changes are made.

Notice also how using variables increases the **readability** of our code. When we use specific values for the pixel location of the ellipse

```
ellipse(75, 100, 40, 40);
```

it is probably not very clear to someone reading our code that we are positioning our circle at the center of the canvas. In contrast, when we use the **width** and **height** variables

```
ellipse(width / 2, height / 2, 40, 40);
```

it is much more likely to be apparent to a reader of our code that the circle is being drawn at the center of the canvas—that is, at the column that is half the width of the canvas and at the row that is half the height of the canvas. The readability of our code is enhanced by the descriptive and meaningful names that Processing gave to these variables, `width` and `height`. Giving descriptive and meaningful names to variables also makes their names easier to remember.

Finally, notice also that we are able to use the `width` and `height` variables more than once in a single program:

```
size(150, 200);
println(width);
println(height);
ellipse(width / 2, height / 2, 40, 40);
```

This illustrates another advantage of a variable: it is **reusable**.

Thus, in exploring some of Processing's built-in variables, we have already discovered several benefits of using variables:

1) Variables can make our programs more *flexible* and easier to modify.

2) Meaningful variable names help to make code more *readable* and easier to understand.

3) Variables are *reusable*.

Creating and Using Our Own Variables

Select **File > New** to start a new program. Save this program as **Variable**.

We have explored two variables that are built into Processing: width and height. However, we also have the ability to create new variables of our own in Processing.

For example, suppose that we would like to draw a square with a side length of 50 pixels in the upper-left corner of the canvas. We can use Processing's rect() function for this. As we learned in Chapter 1, the first two arguments we need to supply to the rect() function specify the column and row where we would like the upper-left corner of our square to be. We know that the upper-left corner of the canvas is at column 0 and row 0. Thus, we can begin by entering the following into the Text Editor area of Processing:

```
rect(0, 0,
```

The next two arguments we need to supply to the rect() function specify the pixel width and pixel height we desire for our rectangle. We wish to draw a square, so we will make these two values the same. Thus, in order to draw a square with a side length of 50, we will complete this statement by adding the following:

```
rect(0, 0, 50, 50);
```

When we run this one-line program, Processing indeed draws a square that is 50 pixels by 50 pixels in the upper-left corner of a canvas that is the default size, 100 pixels by 100 pixels:

Suppose that we would also like to display the side length of our square in the console. For this, we can add the following statement to our program:

```
rect(0, 0, 50, 50);
println(50);
```

When we run our program, in addition to drawing the square on the canvas, Processing also generates the following output to the console:

```
50
```

Now, suppose that we wished to change our square to have a side length of 75 pixels instead. We have used the current side length, 50, in *three* different places in our program. Therefore, we would need to change *all three* of these occurrences of 50 to 75:

```
rect(0, 0, 75, 75);
println(75);
```

Let's *not* make these three changes. Instead, make sure your program still consists of the following:

```
rect(0, 0, 50, 50);
println(50);
```

Fortunately, there is an alternative to making these multiple changes: as we will see, we can modify our program to make use of a *variable* to store the side length of the square.

Save the Variable program. We will continue working with it in an upcoming section.

Variable Names

A variable is a *storage location* that has been given a *name*, and it is important that the name we give each variable is an appropriate one.

In Processing, a variable name must begin with a **letter**. The most common practice is to make this first letter a **lowercase** letter. The characters that follow this first letter can be letters, numbers, or underscores.

Examples of variable names that *can* be used

```
side
side2
side_2
side_length
sideLength
```

Examples of variable names that *cannot* be used

```
7days        Must begin with a letter
side length  Must not contain a space
side-length  Must not contain an operator (-)
```

Variable names should be *meaningful* and *descriptive*. This makes your code more readable and makes the variable names easier to remember. Try to avoid using cryptic or generic variable names such as

```
s
num
```

It is also considered good programming style to avoid using flippant variable names such as

```
whatever
junk
```

When multiple words are desirable for a variable name, a common practice is to combine the words into a single variable name and capitalize the first letter of each word (not including the first word). This practice is sometimes known as **"camel case"** (because the uppercase letters somewhat resemble the humps of a camel). Examples

```
squareSide
sideLength
squareSideLength
```

At the same time, remember that variable names in Processing are *case sensitive*. Thus, if you name your variable

```
circleRadius
```

then you will *not* be able to refer to this variable as

```
circleradius
```

Also, a variable name should not be the same as a word that already has a special meaning in Processing. Such words are sometimes known as **reserved words** or **keywords**. For example, we shouldn't create a variable named width, because Processing already has a variable named width. (Processing usually calls such keywords to our attention by changing their font color when we type them in the Text Editor.)

Variable Types

In addition to giving a variable a name, we must also define its **type**. This specifies the type of *value* that may be stored in this variable. For example, when a variable is of int type, only an int value can be stored in that variable. Or, if a variable is of float type, then only a `float` value can be stored in that variable. Once the type of a variable is specified in a program, it cannot be changed during that program.*

Declaring a Variable

When we define a variable, we specify both its *type* and its *name*. This step is known as **declaring** a variable.

Let's return to our the `Variable` program. Let's create a variable to store an integer that represents the length of the side of the square that we desire. We'll give this `int` variable the name `side`. To declare this variable, we can write the following **declaration statement**:

```
int side;
```

Behind the scenes, this declaration statement has the effect of setting aside an available storage location in the computer's memory that can hold a value of type `int` and that can be referred to using the name `side`. We cannot see this storage location, but we can visualize it as

A variable must be declared in this manner *before* it can be used. This means that our `side` variable can only be used by statements in the portion of the program that comes *after* this declaration. Thus, in this case we will insert the statement declaring our `side` variable at the very beginning of our current program:

```
int side;

rect(0, 0, 50, 50);
println(50);
```

You may have noticed that some orange underlining appears when we insert this declaration statement.

```
int side;
```

Also, an orange message has appeared in the Message Area:

The value of the local variable "side" is not used

This is not an error message. Rather, it is a kind of *warning* message that appears whenever we first declare a variable. Processing is simply calling our attention to the fact that we

---

* Processing uses what is known as *static* typing of variables. This is different from the way that *dynamic* typing of variables works in a language such as Python, where the type of a variable will change to match whatever type of value is assigned to it, even if multiple types of values are assigned to that variable over the course of a single program. We find that students tend to understand dynamic typing more easily *after* learning static typing.

have created a variable named `side` but have not used this variable anywhere else in our program. However, we will soon be making use of this variable, so we can safely *ignore* this warning.*

Save the `Variable` program. We will continue working with it in the next section.

Assigning a Value to a Variable

Currently, our variable is a storage location named `side` that does not yet contain a value. We can visualize the current state of this variable as

Let's try displaying the value of this variable by making the following change to our program:

```
int side;

rect(0, 0, 50, 50);
println(side);
```

The orange underlining and orange message disappear because we are now making use of our `side` variable. However, our reference to the `side` variable in our call to the `println()` function is now underlined in red:

```
println(side);
```

Also, a red error message has appeared in the Message Area:

The local variable "side" may not have been initialized

This error message provides a pretty good description of the problem we are encountering here: we are trying to display the value stored in the `size` variable, but no value has been stored in this variable yet. Whenever we *declare* a variable in Processing, we must *store* a value in this variable *before* this variable can be used. This is known as **initializing** this variable. **Any variable must be *initialized* before it can be *used*.**

We can initialize a variable simply by storing a value in this variable using Processing's **assignment operator**, which is simply the equals sign (=). For example, to store the value 50 in our `side` variable, we simply write

```
int side;

side = 50;
rect(0, 0, 50, 50);
println(side);
```

---

* To turn off these "orange" warnings, select **File > Preferences** and then uncheck the box labeled "Show warnings."

This is an example of what is known as an **assignment statement**, where a *value* on the *right*-hand side of the assignment operator (=) is stored in a *variable* on the *left*-hand side. We can visualize this assignment operation as

$$side = \boxed{50};$$

As a result of this assignment statement, our new `int` variable named `side` now contains a value. We can visualize this initialized variable as

$$int \boxed{\textbf{50}}$$
$$side$$

Notice that the underlining and error message we saw previously have now disappeared.

Remember: The statements in a program are performed *in the order in which they are listed*.

```
int side;

side = 50;
rect(0, 0, 50, 50);
println(side);
```

The first statement in our program creates the variable named `side`.

```
int side;
```

Next, the second statement then initializes this variable by assigning the value 50 to it.

```
side = 50;
```

It is important to notice here that the assignment statement comes *after* the declaration statement. To illustrate this importance, let's we switch the order of these first two statements:

```
side = 50;
int side;

rect(0, 0, 50, 50);
println(side);
```

When we do, Processing immediately underlines "`side`" in the Text Editor:

```
side = 50;
```

Also, an error message appears in the Message Area:

The variable "side" does not exist

This error results because, in the *first* line of the program, we are now trying to store the value 50 in a variable named `side` *before* such a variable has been created by the declaration statement in the *second* line of the program.

Any variable must be *declared* before it can be used to *store* a value. Thus, we need to switch the order of these two statements back to what we had before:

```
int side;

side = 50;
rect(0, 0, 50, 50);
println(side);
```

Now, we are once again properly declaring the variable before we are assigning a value to it. Thus, when we run this program, we see the same square drawn on the canvas as before:

Also, when the last statement of our program is performed,

```
println(side);
```

a *copy* of the int value currently stored in the `side` variable, 50, is retrieved and is used in our call to the `println()` function. We can visualize this as

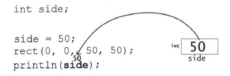

Thus, our call to the `println()` function essentially becomes

```
println(50);
```

As a result of this call to the `println()` function, we now see the value of our `side` variable displayed in the console:

```
50
```

To summarize, we *declared* a variable, then *initialized* this variable by *assigning* a starting value to it, and then *used* this variable in our program.

Save the `Variable` program. We will continue working with it in an upcoming section.

Combining Declaration and Initialization

If we prefer, we are allowed to *combine* declaration and assignment into a single statement. For example, instead of writing

```
int side;

side = 50;
```

we can write

```
int side = 50;
```

This single statement accomplishes two operations: it both *declares* the variable `side` and *initializes* it with the value 50.

There are compelling reasons in favor of either style. However, in this book, we will usually declare and initialize a variable using *separate* statements. This is to help us remember that declaring a variable and initializing a variable are actually two distinct actions. Furthermore, there are some situations in Processing where the combined form *cannot* be used.

Reusing a Variable

Let's return to our `Variable` program. Now that our `int` variable `side` has been declared and initialized, it can be used *multiple* times in place of other specific `int` values within any statements that follow in our program.

For example, we can insert our `int` variable `side` into the two other places in our program where we previously used the specific `int` value, 50:

```
int side;

side = 50;
rect(0, 0, side, side);
println(side);
```

Now, when we run our revised program, each time there is a reference to the `side` variable, a copy of the value currently stored in `side` is retrieved and used in that place. We can visualize this as

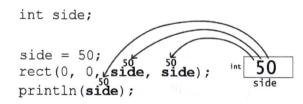

Thus, the call to the `rect()` function

```
rect(0, 0, side, side);
```

essentially becomes

```
rect(0, 0, 50, 50);
```

For this reason, a square that is the same size as before is drawn on the canvas:

Likewise, the call to the `println()` function

```
println(side);
```

once again essentially becomes

```
println(50);
```

Thus, the same side length as before is displayed in the console:

```
50
```

We see here that one of the advantages of a variable is that it is *reusable*. Once the `side` variable was declared and then initialized with a value, we were able to use the `side` variable in three places: twice in our call to the `rect()` function and once in our call to the `println()` function.

We also see here once again how using a variable also improves the readability of our programs. The purpose of the unexplained specific values in these statements isn't very clear:

```
rect(0, 0, 50, 50);
println(50);
```

In contrast, the use of the variable name `side` in these two statements makes our program easier to understand:

```
int side;

side = 50;
rect(0, 0, side, side);
println(side);
```

The term *variable* also suggests *variation*. Indeed, one of the reasons for creating and using a variable is that we might wish to *change* the value stored in the variable.

For example, if we decide to change the side length of our square to 75 pixels, we only need to make *one* simple change to the assignment statement:

```
int side;

side = 75;
rect(0, 0, side, side);
println(side);
```

Now, when we run our program, wherever there is a reference to the `side` variable, a copy of the `int` value currently stored in `side` is retrieved, 75, and is used in that place. We can visualize this as

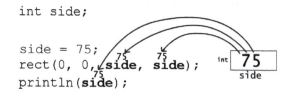

Thus, the call to the `rect()` function

```
rect(0, 0, side, side);
```

essentially becomes the following:

```
rect(0, 0, 75, 75);
```

Likewise, the call to the `println()` function

```
println(side);
```

essentially becomes

```
println(75);
```

Thus, after this single change to our program, a square with a side length of 75 pixels is now drawn:

This new side length is also displayed in the console:

```
75
```

By using variables, we make it easier to make any necessary changes than if we use specific values.

Save the `Variable` program. We will keep working with it in the next section.

Type Mismatches

Return to the `Variable` program.

Remember: Every variable is declared to be of a specific *type*. Our `side` variable is currently declared to be of type `int`:

```
int side;
```

We can visualize this variable as

Declaring `side` to be an `int` variable means that the name `side` now refers to a storage location that can only store an `int` value. Our assignment statement initializing this variable,

```
side = 75;
```

is permitted because both the *value* on the *right*-hand side of the assignment operator (=) and the *variable* on the *left*-hand side are of type `int`. We can visualize this assignment operation as

And we can visualize the result of this assignment statement as

However, suppose we try changing our assignment statement to

```
int side;

side = 75.0;
rect(0, 0, side, side);
println(side);
```

Processing now underlines a portion of our assignment statement:

```
side = 75.0;
```

Also, we see the following error message in the Message Area:

Type mismatch, "float" does not match with int

As this error message indicates, we now have what is known as a **type mismatch**.

Remember, in an assignment statement, the *value* on the *right*-hand side of the assignment operator (=) is stored in the *variable* on the *left*-hand side. Only an int value may be assigned to an int variable. Here, we have tried to store the float value 75.0 in our int variable side. We can visualize the type mismatch in this assignment operation as

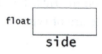

To correct this error, let's change the declared type of the size variable to be **float** instead:

```
float side;

side = 75.0;
rect(0, 0, side, side);
println(side);
```

The underlining and the error message now disappear. This is because the variable name side now refers to a location in memory that can only store a value of type float, which we can visualize as

float ▢
side

Our assignment statement initializing this variable is now permitted because both the value on the right-hand side of the assignment operator (=) and the variable on the left-hand side are of type float. We can now visualize this assignment operation as

side = 75.0;
(float)   (float)

And we can visualize the result of this assignment statement as

float 75.0
side

Now, when we run our program, wherever there is a reference to the side variable, a copy of the float value currently stored in side is retrieved and used in that place. We can visualize this as

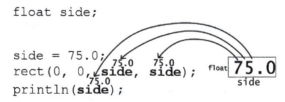

Thus, the call to the rect() function

```
rect(0, 0, side, side);
```

essentially becomes the following:

```
rect(0, 0, 75.0, 75.0);
```

As it turns out, almost all of Processing's *drawing* functions allow us to supply `float` values as arguments. Thus, we may also use `float` values in our call to the `rect()` function. As a result, a 75-pixel by 75-pixel square is once again drawn on the canvas:

Likewise, the call to the `println()` function

```
println(side);
```

essentially becomes the following:

```
println(75.0);
```

Thus, the side length is now displayed in the console as a `float` value:

```
75.0
```

Not all type mismatches cause error messages. For example, let's now change the value on the right-hand side back to an `int` value.

```
float side;

side = 75;
rect(0, 0, side, side);
println(side);
```

This creates another type mismatch, because the value on the right-hand side of the assignment operator is an `int` value, but the variable on the left-hand side is of type `float`. We can visualize the type mismatch in this assignment operation as

$$\text{side} \;=\; \boxed{75}\text{;}$$
$$\text{\small(float)}\qquad\qquad\text{\small(int)}$$

However, this type mismatch does *not* produce underlining or an error message. In fact, if we go ahead and run our program, we see that the square drawn on the canvas has the same 75-pixel side length as before:

We also still see the same float value for the side length output to the console:

```
75.0
```

Why are we seeing a float value output to the console when we assigned an int value the side variable? Here's what happened: Processing detected a type mismatch when we attempted to assign an int value (75) to a float variable (side). However, in such cases, Processing *automatically* resolves the type mismatch by *converting* the int value to the equivalent float value. We can think of this as Processing automatically appending " .0 " to the end of the int value, making it a float value. Thus, in this particular program, Processing automatically converts the int value 75 to the float value 75.0 and then assigns this float value to the float variable side. We can visualize the automatic conversion in this assignment operation as

$$
\overset{\overset{\displaystyle 75.0}{\text{\tiny (float)}}}{\underset{\text{\tiny (float)}}{\texttt{side}} \;=\; \underset{\text{\tiny (int)}}{\boxed{75}};}
$$

And we can visualize the result of this assignment statement as

$$
\texttt{float}\boxed{\textbf{75.0}}
$$
$$
\texttt{side}
$$

This automatic conversion from int to float is why, when we run our program, we do not receive an error message as a result of the type mismatch in our assignment statement, and why we see a float value output to the console instead of the int value that we used in our assignment statement:

```
75.0
```

In summary, it is important to remember the following about type mismatches:

- Once a variable is declared, *only a value of the defined type may be stored in that variable.*

Thus,

- If we attempt to assign a float value to an int variable, then an error message will result.

int ✖ float

- If we attempt to assign an int value to a float variable, the int value is automatically converted to a float value. (We can think of this as simply the appending of .0 onto the end of the int value.)

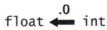

You might more easily remember how Processing handles such a type mismatch by thinking of it in the following way:

- Processing *will not* make the mismatched value **"shorter."** That is, it *will not* "cut off" the fractional portion of a float value to make an int value.

- Processing *will* make the mismatched value **"longer."** That is, it *will* essentially "tack on" a .0 to an int value in order to make it a float value.

$$75 \longrightarrow 75.0$$

Save the Variable program. We will continue working with it in the next section.

Why Not Use Only the float Type?

Because of the automatic type conversion from int to float that Processing does, it might be tempting to declare all numeric variables to be of type float. However, here are some reasons why you should *not* do this:

1) It is important to understand the key differences between the int and float types, such as the difference between the way *division* is performed with int values and the way division is performed with float values.

2) Any int value is always stored exactly in an int variable, whereas float fractions are sometimes inaccurate.

3) There are certain situations in Processing where only int values may be used.

These are only some of the reasons why, in general, you should *avoid* declaring a variable to be of type float when you expect that all the values to be stored in this variable will be integers.

For example, although Processing allows us to use float values with most of its drawing functions, any fractional portion of a float value used with a drawing function will usually be ignored because Processing only works with *whole* pixels. Thus, a function call such as

```
rect(0.5, 0.5, 50.2, 50.2);
```

will essentially be treated as

```
rect(0, 0, 50, 50);
```

For this reason, it makes more sense to make a habit of using `int` values with Processing's drawing functions. Thus, in our current program, let's change the type declaration of our `side` variable back to `int`.

```
int side;

side = 75;
rect(0, 0, side, side);
println(side);
```

Save the `Variable` program. We will keep working with this program in the next section.

Expressions in Assignment Statements

Return to the `Variable` program. Select **File > Save As** and resave this program as **MoreAssigning**.

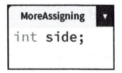

The `int` or `float` result of an arithmetic *expression* may also be assigned to a variable of the same type. For example, suppose that we would like to have the side length of our square be *one-fourth* of the width of the canvas. We know that the width of the canvas is 100 pixels by default.

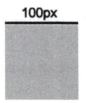

Thus, knowing that 100 ÷ 4 equals 25, we could simply use this specific value as the side length of our square:

```
int side;

side = 25;
rect(0, 0, side, side);
println(side);
```

However, another way that we could achieve our desired result is to write the following and have Processing perform the calculation for us:

```
int side;
```

```
side = 100 / 4;
rect(0, 0, side, side);
println(side);
```

We now have an assignment statement with an arithmetic *expression* on the right-hand side of the assignment operator (=):

```
side = 100 / 4;
```

It is important to understand that whenever an assignment statement contains an expression:

1) The expression on the *right-hand* side of the assignment operator (=) will always be evaluated *first*.

2) After the evaluation of the expression is complete, the value that is the *result* of this evaluation is assigned to the variable on the *left-hand* side.

For example, in the current assignment statement, the expression on the *right-hand* side is evaluated *first*.

$$side = \boxed{100 \ / \ 4};$$

This particular expression evaluates to the int value 25:

$$side = \overset{\textbf{25}}{\boxed{100 \ / \ 4}};$$

The evaluation of the right-hand side is complete. We now have a single int value on the right-hand side. Thus, our assignment statement has essentially become

$$side = \boxed{\textbf{25}};$$

Both the variable on the left-hand side of the assignment operator and the value on the right-hand side are of type int.

$$\underset{\text{(int)}}{side} = \underset{\text{(int)}}{\boxed{\textbf{25}}};$$

This means that this assignment statement will successfully store the int result from the right-hand side, 25, in the int variable on the left-hand side of the assignment operator, side. We can visualize this assignment operation as

$$\underset{\text{(int)}}{side} = \underset{\text{(int)}}{\boxed{\textbf{25}}};$$

And we can visualize the result of this assignment statement as

Indeed, when we run our program, we see the following square drawn on the canvas,

and we can see from the output to the console that this square has a side length of 25 pixels:

    25

Now, suppose we try modifying our assignment statement so that it instead contains the following expression:

```
int side;

side = 100 * 0.25;
rect(0, 0, side, side);
println(side);
```

In our modified assignment statement, one of the operands, 0.25, is now of type float.

$$side = \underset{(int)}{100} * \underset{(float)}{0.25};$$

Remember: If *either* or *both* of the operands in an arithmetic expression are of type float, then the *result* of that arithmetic operation is of type float. Thus, the expression on the right-hand side evaluates to 25.0, a float value. Thus, this assignment statement essentially becomes

$$\underset{(int)}{side} = \underset{(float)}{25.0};$$

We are trying to assign a float value to an int variable. For this reason, the expression is now underlined:

    side = 100 * 0.25;

And, once again, we receive the following error message:

Type mismatch, "float" does not match with "int"

We have a type mismatch because the resulting float value, 25.0, cannot be stored in the int variable side. We can visualize the type mismatch in this assignment operation as

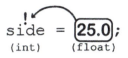

An expression on the *right*-hand side of an assignment statement must evaluate to a value that is compatible with the type of the variable on the *left*-hand side. For this reason, we will change the expression in this statement back to an expression that we know evaluates to an `int` value, eliminating the type mismatch:

```
int side;

side = 100 / 4;
rect(0, 0, side, side);
println(side);
```

Save the `MoreAssigning` program. We will continue working with it in the next section.

Using a Variable on the Right-Hand Side of an Assignment Statement

Return to the `MoreAssigning` program. Currently we are using a specific `int` value, `100`, to represent the width of our canvas:

$$\text{side} = \boxed{100} / \ 4;$$

However, we also know that Processing always automatically stores the width of the current canvas in its predefined `width` variable, which is also of type `int`. In this particular case, we can visualize this initialized as

$$\text{int} \ \boxed{\textbf{100}}$$
$$\text{width}$$

Thus, we could improve our current example by replacing the specific value, `100`, with a reference to Processing's predefined `width` variable:

```
int side;

side = width / 4;
rect(0, 0, side, side);
println(side);
```

When this new version of our assignment statement is performed, as always, the expression on the *right-hand* side is evaluated *first*:

$$\text{side} = \boxed{\textbf{width} \ / \ 4};$$

This particular expression contains a reference to a variable, `width`. Whenever a reference is made to a variable in a Processing statement, a *copy* of the value that is currently stored in this variable is retrieved from the variable.

Thus, the process of evaluating this particular expression begins by retrieving a copy of the `int` value stored in the `width` variable. We can visualize this as

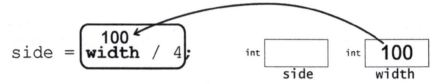

This value retrieved from the variable is then used in the division operation

$$side \ = \boxed{100 \ / \ \ 4};$$

and the evaluation of the expression on the right-hand side can now be completed.

$$\overset{25}{side \ = \boxed{100 \ / \ \ 4}};$$

The evaluation of the right-hand side is complete. Thus, the assignment statement essentially becomes

$$side \ = \boxed{25};$$

In this particular case, the value that results, 25, is of type `int`. The `side` variable is also of type `int`.

$$\underset{(int)}{side} \ = \boxed{\underset{(int)}{25}};$$

Thus, there is no type mismatch. The `int` result of the evaluated expression on the right-hand side can indeed be stored in the `int` variable on the left-hand side. We can visualize this assignment operation as

$$\underset{(int)}{side} \ = \boxed{\underset{(int)}{25}};$$

The performance of the assignment statement is now complete. We can visualize the result of this completed assignment as

$$\text{int} \ \boxed{\begin{array}{c} 25 \\ side \end{array}}$$

Thus, when we run this new version of our program, we do indeed see a square with a side length of 25 pixels drawn on the canvas, and the value of the side variable that is output to the console is indeed 25:

25

Using the width variable to set our side variable makes this program more flexible. The side length of our square will always be equal to one-fourth of the width of the canvas, no matter what size we choose for our canvas with the size() function.

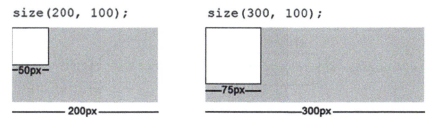

Now, suppose that we would like the side length of our square to be equal to exactly *three-fourths* of the width of the canvas. One way to accomplish this result is by making the following change to the assignment statement in our current program:

```
int side;
side = width * 3 / 4;

rect(0, 0, side, side);
println(side);
```

The default canvas width is 100 pixels. Three-fourths of this canvas width is 75. When we run this program, we do indeed see a square drawn that with a side length looks to be equal to three-fourths of the width of the canvas:

We also see the expected side length, 75, displayed in the console:

75

Let's take a closer look at our current assignment statement so that we can better understand our resulting side length. As always, the *right-hand* side of the assignment statement is evaluated *first*.

$$\texttt{side} = \boxed{\texttt{width * 3 / 4}};$$

The right-hand side of this particular assignment statement consists of an arithmetic expression containing *two* arithmetic operations: multiplication and division. Which operation is performed first? According to the rules of the order of operations, multiplication and division have the same level of precedence, so this expression is simply evaluated from left to right. Thus, the multiplication operation is performed first:

$$\text{side} = \boxed{\text{width} * 3} / 4;$$

For this operation, a copy of the value stored in the width variable, 100, is retrieved. We can visualize this as

$$100 \leftarrow \text{side} = \boxed{\text{width} * 3} / 4; \qquad \text{int} \boxed{100}_{\text{width}}$$

Thus, the multiplication operation becomes

$$\text{side} = \boxed{100 * 3} / 4;$$

When this multiplication operation is performed, the expression on the right-hand side now becomes essentially

$$\text{side} = \boxed{300 / 4;}$$

This remaining division operation is performed next. This is an integer division operation, since both operands are int values. Thus, the result of this division operation is an int value, 75. Thus, our assignment statement becomes essentially

$$\text{side} = \boxed{75};$$

The evaluation of the right-hand side is now complete. The resulting value, 75, is an int value. Thus, it can be stored in the int variable side. We can visualize this assignment operation as

$$\text{side}_{\text{(int)}} = \boxed{75}_{\text{(int)}};$$

The performance of the assignment statement is now complete. And we can visualize the result as

$$\text{int} \boxed{75}_{\text{side}}$$

This side length, 75, is indeed equal to three-fourths of the width of the canvas, 100. In fact, no matter what size canvas we choose, the side length of the square will always be three-fourths of the width of the canvas.

As we have seen in the two preceding examples, an assignment statement can contain an *expression* on the right-hand side of the assignment operator (=). In such cases, Processing always *evaluates* this expression *first*. If there is not a type mismatch, Processing then assigns the resulting value to the variable on the left-hand side.

Save the MoreAssigning program. We will continue working with it in the next section.

Being Careful with Integer Division

Select **File > Save As** and resave the MoreAssigning program as **Division**.

Earlier in this chapter, the use of parentheses was recommended whenever an expression contains two or more arithmetic operations, both for clarity and to remove the need to rely solely upon the rules for the order of operations. Thus, you may have wondered whether we should have added parentheses to the assignment statement in our current program for the purpose of improving readability. Specifically, in order to clarify that we are using three-fourths of the canvas width as the side length of our square, we might try inserting parentheses, as in the following:

```
int side;
side = width * (3 / 4);

rect(0, 0, side, side);
println(side);
```

However, if we make this change, then when we run this program, we now see no square on the canvas:

We also see that the side length of our square displayed in the console is now zero:

0

Having a side length of zero certainly explains the absence of the square on the canvas. However, why is zero our resulting side length?

The rules governing the order in which arithmetic operations are performed specify that parentheses have a higher priority than multiplication. Thus, when we add parentheses to the expression

```
side = width * (3 / 4);
```

the division operation contained within the parentheses will now be performed *before* the multiplication.

$$\text{side } = \text{ width } * \boxed{(3\ /\ 4)};$$

We might have been hoping that the result of this division operation would be

```
0.75
```

However, the operands of this division operation, 3 and 4, are both of type int. Remember what we have learned about division with integers in Processing: the resulting quotient is always an *integer*. There is no fractional portion in the quotient. Thus, the result of the integer division operation 3 / 4 is simply zero. We can visualize the result of this division operation as

$$\text{side } = \text{ width } * \overset{\textbf{0}}{\boxed{(3\ /\ 4)}};$$

The multiplication operation then becomes

$$\text{side } = \boxed{\text{width } * \ \textbf{0}};$$

No matter what int value is stored in the width variable, this multiplication operation always produces zero as a result. Thus, the assignment statement always becomes

$$\text{side } = \boxed{\textbf{0}}\ ;$$

The evaluation of the right-hand side is complete, so the resulting value, 0, can now be stored in the side variable. We can visualize this assignment operation as

$$\text{side } = \boxed{\textbf{0}}\ ;$$

And we can visualize the result of this assignment operation as

$$\overset{\text{int}}{}\boxed{\textbf{0}}$$
$$\text{side}$$

Notice that running our program did not produce an error message. However, we did not obtain the result we desired. Such an error is known as a **logic error**. How can we fix this logic error? One option is to remove the parentheses and return the statement to what we had before:

```
side = width * 3 / 4;
```

Another option is to eliminate the integer division by changing the operands to float values:

```
side = width * (3.0 / 4.0);
```

However, when we make this change, underlining appears under the entire expression:

```
side = width * (3.0 / 4.0);
```

Also, an error message appears in the Message Area:

Type mismatch, "float" does not match with "int"

Can you guess why we are receiving this underlining and error message? The division operation with `float` operands does eliminate the problem of integer division, because it produces a result that is a `float` value:

0.75

$$\text{side = width *}\ \boxed{\text{(3.0 / 4.0)}};$$

Thus, when the value in the `width` variable is retrieved,

100

$$\text{side = width * 0.75;}\qquad \text{int}\ \boxed{\textbf{100}}$$
width

the multiplication operation becomes

$$\text{side =}\ \boxed{\textbf{100 * 0.75}}\ ;$$

However, remember that whenever even *one* of the operands is of type `float`, the result of that arithmetic operation is also of type `float`. Thus, the current multiplication operation produces a `float` value, 75 .0, as its result, so our assignment statement essentially becomes

$$\text{side =}\ \boxed{\textbf{75.0}}\ ;$$

We now have a type mismatch. The resulting `float` value, 75 .0, cannot be stored in `side`, a variable of type `int`. We can visualize the type mismatch in this assignment operation as

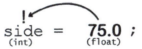

$$\underset{\text{(int)}}{\text{side}} = \underset{\text{(float)}}{\textbf{75.0}}\ ;$$

One way that we can resolve this type mismatch is by changing the declared type of the `side` variable from `int` to `float` :

```
float side;
side = width * (3.0 / 4.0);

rect(0, 0, side, side);
println(side);
```

Now, `side` is a variable that can store a `float` value. We can visualize this variable as

float ⎿_____⎽
side

This change eliminates the type mismatch. The underlining and error message disappear. Thus, 75.0, a float value, can now be stored in the float variable side. We can visualize this assignment operation as

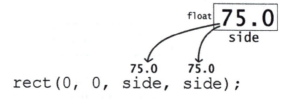

And we can visualize the result of this assignment operation as

float `75.0`
side

It was previously noted that most of Processing's drawing functions permit us to use float values as arguments when we call them, and the rect() function is no exception. Even though side is now a float variable, we can use it as an argument in our call to the rect() function. When the rect() function is called, copies of the float value stored in side, 75.0, are retrieved. We can visualize this as

Thus, the call to the rect() function essentially becomes

```
rect(0, 0, 75.0, 75.0);
```

Indeed, when we run our program, we see that the same square as before is drawn on the canvas:

The console output also shows that the value of the side variable is now a float value:

```
75.0
```

As we have seen in this example, making one or both of the operands type float can often help us avoid problems that result from integer division.
    Save and close the Division program.

Reassigning a New Value to a Variable

Return to the Variable program. Select **File > Save As** and resave the Variable program as **Reassign**.

Our current program is

```
int side;

side = 75;
rect(0, 0, side, side);
println(side);
```

When we run this program, we see on the canvas a square with a side length of 75 pixels.

The console output

75

also demonstrates that the value currently stored in our `size` variable is 75. We can visualize this variable as

$$\text{int}\ \boxed{\textbf{75}}$$
$$\text{side}$$

Once we have declared, initialized, and used a variable, we are also able to *reassign* a new value to our variable as many times as we would like to do so within a single program. To illustrate, let's make some additions to our program so that, after drawing the current square that has a side length of 75 pixels, it then draws a second square that has a smaller side length, 50 pixels:

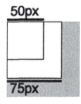

As we'll see, we are able to draw this second square with a side length of 50 pixels simply by *reassigning* a new `int` value, 50, to our `side` variable and then *reusing* the `side` variable in our calls to the `rect()` and `println()` functions.

In order to assign a new value to `side`, do we need to *declare* this variable again? Let's try inserting a second declaration statement for our `side` variable:

```
int side;

side = 75;
rect(0, 0, side, side);
println(side);

int side;
```

Immediately, an underline appears under this second declaration of a `side` variable:

```
int side;
```

Also, the following error message appears in the Message Area:

Duplicate local variable side

We are attempting here to declare a second variable with the name `side`, and this is not permitted. We cannot have two variables with the same name in this program.*

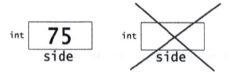

The fact is, we do not need to declare `side` a second time. As the term *variable* suggests, the values of variables are allowed to vary. Thus, what we want to do here is simply *reassign* a new value to our `side` variable so that we can then *reuse* our `side` variable to draw the second square. We'll *reassign* a new value to `side` simply by adding the following assignment statement at the end of our current program:

```
int side;

side = 75;
rect(0, 0, side, side);
println(side);

side = 50;
```

This assignment has the effect of replacing the `int` value that was stored in our `size` variable, 75, with the `int` value 50. We can visualize this change as

---

* When we learn later how to divide our programs into multiple blocks and functions, it will technically be possible to have duplicate variable names in the same program. However, this can be confusing, so we will usually give variables distinct names.

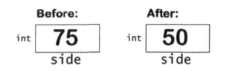

Thus, at this point in the program, the value stored in our `side` variable is changed from 75 to 50. This means that we can now reuse `side` to draw the second square and output the new value of `side` to the console simply by copying the calls to the `rect()` function and `println()` function that we used for the first square:

```
int side;

side = 75;
rect(0, 0, side, side);
println(side);

side = 50;
rect(0, 0, side, side);
println(side);
```

Indeed, when we run our program, we now see a second square drawn on top of the first square,

and our console output of the two values of the two side lengths is now

```
75
50
```

How did our changes achieve these results? To clarify this, let's step through our program, one statement at a time. This is known as tracing the **flow of execution**.

It is important to remember that the statements of a program are performed in *sequence*, the order in which they are listed in our program.

```
int side;

side = 75;
rect(0, 0, side, side);
println(side);

side = 50;
rect(0, 0, side, side);
println(side);
```

When we run our program, the first statement listed is the first statement performed:

```
int side;
```

This declaration statement has the behind-the-scenes effect of setting aside a storage location in the computer's memory that can hold an int value and can be referred to with the name "side." We can visualize the result of declaring this variable as

side

Next, the second statement is performed:

```
int side;

side = 75;
```

This second statement is an assignment statement. It has the effect of initializing our variable by storing the int value 75 in our side variable. We can visualize the result of this assignment statement as

int **75**

side

When the third statement is performed,

```
int side;

side = 75;
rect(0, 0, side, side);
```

the rect() function is called using the side variable as the two arguments specifying the width and height of the rectangle, respectively. A copy of the value currently stored in the side variable, 75, is retrieved for each of these two references to the side variable in the call to the rect() function. We can visualize this as

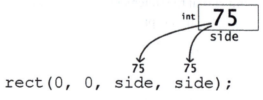

Thus, this call to the rect() function essentially becomes

```
rect(0, 0, 75, 75);
```

When this third statement is performed using these retrieved values of the side variable, the first square is drawn:

The fourth statement,

```
int side;

side = 75;
rect(0, 0, side, side);
println(side);
```

also contains a reference to the side variable. Thus, a copy of the value currently stored in the side variable, 75, is retrieved. We can visualize this as

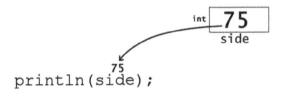

Thus, this call to the println() function essentially becomes

```
println(75);
```

As a result, this value is displayed in the console:

75

The fifth statement *reassigns* a new int value, 50, to our side variable:

```
int side;

side = 75;
rect(0, 0, side, side);
println(side);

side = 50;
```

We can visualize the result of this assignment statement as

When the sixth statement is performed,

```
int side;

side = 75;
rect(0, 0, side, side);
println(side);

side = 50;
rect(0, 0, side, side);
```

the `rect()` function is once again called using the `side` variable as the arguments specifying the rectangle's width and height, respectively. A copy of the value currently stored in
the `side` variable, 50, is retrieved for each of these two references to the `side` variable in
the call to the `rect()` function. We can visualize this as

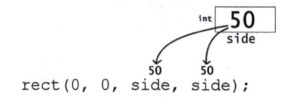

Thus, this call to the `rect()` function essentially becomes

```
rect(0, 0, 50, 50);
```

When this sixth statement is performed using these retrieved values of the `side` variable, the second square is drawn. This second smaller square is drawn on top of the first,
larger, 75-pixel-sided square because, according to Processing's stacking order, the *last*
element drawn on the canvas is always drawn on *top* of any elements that are already on
the canvas.

The seventh and final statement,

```
int side;

side = 75;
rect(0, 0, side, side);
println(side);

side = 50;
rect(0, 0, side, side);
println(side);
```

also contains a reference to the side variable. Thus, a copy of the value currently stored in the side variable, 50, is retrieved. We can visualize this as

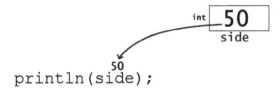

```
                    50
      println(side);
```

This call to the println() function essentially becomes

```
println(50);
```

As a result, this second value is also displayed in the console:

```
50
```

Thus, the total console output of this program is

```
75
50
```

This completes our trace of the sequential performance of all the statements in our program.

A single variable can be reassigned and reused as many times as we would like. For example, if we would like to add a third, even smaller square with a side length of 25 pixels, we simply reassign 25 to our side variable and once again copy the calls to rect() and println() functions that we used for the other two squares:

```
int side;

side = 75;
rect(0, 0, side, side);
println(side);

side = 50;
rect(0, 0, side, side);
println(side);

side = 25;
rect(0, 0, side, side);
println(side);
```

When we run this program, we now see three squares drawn on the canvas:

We also see their three side lengths displayed in the console:

```
75
50
25
```

Save and close the `Reassign` program.

Constants

Select **File > New** to create a new program. Save this program as **Constants**.

A **constant** is similar to a variable in that it is a named storage location that stores a value of a certain type. However, unlike the value stored in a variable, the value stored in a constant **cannot be changed**. Also, constants are typically given names that are *all capital letters* to help distinguish them from variables.

Predefined Constants

Similar to Processing's predefined *variables* (for example, `width` and `height`), there are also a number of predefined *constants* built into Processing.

A good example is Processing's predefined `float` constant named `PI`. This constant has been preassigned a value by Processing: the mathematical number known as pi (π) to seven decimal places. To illustrate, if we write the following one-line program to display `PI` to the console,

```
println(PI);
```

we see the following output:

```
3.1415927
```

Notice that the name of this constant, `PI`, is comprised entirely of capital letters, as is customary for constants.

We can use the `PI` constant in some of the same ways that we would use a variable. For example, the formula for the circumference of a circle is pi multiplied by the diameter. Thus, to display the circumference of a circle with a diameter of 20 pixels, we might write the following one-line program:

```
println(PI * 20);
```

This statement causes the following `float` value to be displayed in the console:

```
62.831856
```

However, if we try to modify the value assigned to the PI constant,

`PI = 3.14;`

then underlining appears in this statement,

`PI = 3.14;`

and we see an error message indicating that the constant PI cannot be used in an assignment statement. This underlining and error message appears because the constant PI already has a value assigned to it by Processing. Once a constant has had a value assigned to it, no other value can be assigned to this constant. This illustrates the difference between a constant and a variable: the value of a variable can vary during a program; the value of a constant cannot.

Other predefined constants built into Processing include QUARTER _ PI, HALF _ PI, and TWO _ PI. Along with PI, these constants can be used in the arc() function to draw certain arcs without having to convert to radians, because each of these constants is already assigned the measure in radians of the angle that its name describes:

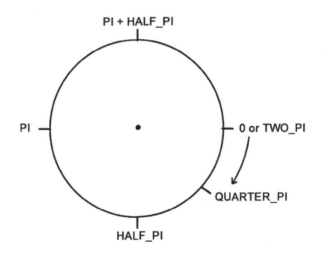

Recall, the first four arguments of the arc() function are the same as those of the ellipse() function, specifying the center and size of the ellipse. The additional two arguments in the arc() function specify the start angle and stop angle of the arc to be drawn. These angle measures must be given in *radians*. Previously, we have used Processing's radians() function to specify these arguments. However, we can also use these PI, QUARTER _ PI, HALF _ PI, and TWO _ PI constants to refer to the angles that their names describe. For example,

`arc(width / 2, height / 2, 80, 80, `**`0, radians(45)`**` );`

can be written instead as

```
arc(width / 2, height / 2, 80, 80, 0, QUARTER_PI);
```

and will draw the following arc:

Processing also defines several constants of type int that can be used to select a specific mode for certain of its functions. For example, as we have seen, the default drawing mode for the arc() function is a pie-shaped wedge without any line segments connecting the two angle positions. However, there are three other drawing modes for the arc() function. These numbered modes can be selected using the three predefined and descriptively named constants: CHORD, OPEN, and PIE.

For example, if we use the PIE constant as an added argument in the arc() function, then the arc is instead drawn as a pie wedge that has line segments connecting the two angle positions. For example,

```
arc(width / 2, height / 2, 80, 80, 0, QUARTER_PI, PIE);
```

will draw the following arc:

Having this predefined constant works nicely for specifying this mode, much nicer than having to specify this mode using a specific integer.

Information about the CHORD and OPEN constants and the drawing modes they select is available in Processing's Reference entry for the arc() function.

Delete all the statements that currently comprise the Constants program. We will compose new statements for this program in the next section.

Defining Our Own Constants

Return to the Constants program. There may be occasions when we would like to define our own constants. To do this, we simply add the keyword **final** to the variable declaration and *capitalize* the variable name.

For example, if we wanted our variable storing the side length of the square to be a constant so that it could not be modified later in the program after it is initialized, we could write the following statement to declare such a constant:

```
final int SIDE;
```

We can visualize the result of this statement as

We can then initialize this constant by adding an assignment statement:

```
final int SIDE;
SIDE = 50;
```

However, more commonly, a constant is both declared and initialized in a single statement, as in the following:

```
final int SIDE = 50;
```

Once a constant is initialized, the value it contains cannot be changed. We can visualize this as

A constant can be used in much the same way that we use a variable. For example, to use the constant SIDE in a program, we can write

```
final int SIDE = 50;

rect(0, 0, SIDE, SIDE);
println(SIDE);
```

When we run this program, we see on the canvas a square that has a side length equal to our SIDE constant,

and we also see the value of our SIDE constant output to the console:

```
50
```

However, if we try to modify the SIDE constant after it has been initialized,

```
final int SIDE = 50;

SIDE = 25;
```

```
rect(0, 0, SIDE, SIDE);
println(SIDE);
```

then underlining appears in this added statement,

```
SIDE = 25;
```

and we see an error message that begins with

The final local variable SIDE cannot be assigned.

We will not use constants often in this book, but it is important to be able to recognize constants when they are used and to understand the rules for using them.

Save and close the Constants program.

Nonnumeric Types

Part of what sets Processing apart from other languages for beginners is its many functions dedicated to graphics. However, we have seen in this chapter that we are also able to do numeric operations in Processing.

As we shall see in the next sections, Processing also provides us with the ability to work with *non*numeric information, such as individual characters and strings of multiple characters.

Individual Characters: The char Type

Select **File > New** to start a new program. Save this program as **CharType**.

Processing provides the **char** type for creating a variable that can store a single character. For example, such a character can be a letter of the alphabet, a numeral from 0 to 9, or a punctuation mark.

A *value* that is of type char is enclosed in a pair of *single* quotes. Thus, all the following are examples of char values:

```
'a'
'A'
'9'
'!'
'@'
```

Remember, a variable is actually a named storage location in the computer's memory where we can store a specific *type* of value. To declare a *variable* of type char, we simply use a declaration statement of the same form as that which we have used for numeric variables. For example, the following statement creates a variable that can hold a single character:

```
char letter;
```

We can visualize this variable as

char | [   ]
letter

To store a value of type `char` in this variable `letter`, we simply use an assignment statement of the same form as that which we have used for numeric variables. However, we must remember to enclose the letter, numeral, or punctuation mark in a pair of single quotes when assigning this `char` value to our variable. For example, adding the following assignment statement stores a lowercase r in the `letter` variable:

```
char letter;
letter = 'r';
```

We can visualize the result of this declaration and initialization as

char | 'r'
letter

Behind the scenes, Processing's `char` type uses the **Unicode** coding system, which assigns a unique integer to each letter, numeral, and punctuation mark in a wide variety of languages. For example, the decimal versions of the Unicode numbers for the uppercase letters of the English alphabet are

A	B	C	D	E	F	G	H	I	J	K	L	M
65	66	67	68	69	70	71	72	73	74	75	76	77

N	O	P	Q	R	S	T	U	V	W	X	Y	Z
78	79	80	81	82	83	84	85	86	87	88	89	90

Interestingly, each lowercase letter of the English alphabet has a decimal Unicode number that is 32 higher than that of its corresponding uppercase letter:

a	b	c	d	e	f	g	h	i	j	k	l	m
97	98	99	100	101	102	103	104	105	106	107	108	109

n	o	p	q	r	s	t	u	v	w	x	y	z
110	111	112	113	114	115	116	117	118	119	120	121	122

For example, the lowercase r character is assigned the decimal Unicode number 114. Behind the scenes, Processing uses this integer to store the a lowercase r in a `char` variable. Thus, we might also visualize the result of declaring and initializing the `letter` variable as

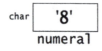

We can output a value stored in this char variable using the println() method in the same manner as we have done before:

```
char letter;
letter = 'r';
println(letter);
```

When we run this program, we see the following output to the console:

r

A **numeral** can also be stored in a char variable. The numerals 0–9 have the following Unicode numbers:

0	1	2	3	4	5	6	7	8	9
48	49	50	51	52	53	54	55	56	57

For example, the following program creates a char variable named numeral and stores the numeral 8 in it:

```
char numeral;
numeral = '8';
```

The numeral 8 is assigned the decimal Unicode number 56. Behind the scenes, Processing uses this integer to store the numeral 8 in a char variable. Thus, we can visualize the result of this program as

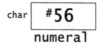

or as

If we add the following call to the println() function to our program,

```
char numeral;
numeral = '8';
println(numeral);
```

then this stored character is displayed in the console:

8

It is important not to confuse the use of a `char` variable,

```
char numeral;
numeral = '8';
```

with the use of an `int` variable:

```
int number;
number = 8;
```

Because a `char` value is stored internally as an integer according to the Unicode number-ing system, using a `char` value or variable in an arithmetic expression will not produce an error. However, it will produce a result that is different from what we might expect. For example, the following program,

```
char numeral;
numeral = '8';
println(numeral + 1);
```

gives the following output to the console:

```
57
```

This output results because the numeral 8 has the decimal Unicode character number 56. Thus, when 1 is added to the `char` variable `numeral`, 57 is the result.

A **punctuation mark** or other **keyboard symbol** can also be stored in a `char` variable. For example, the following program creates a `char` variable named `atSign` and stores the at sign in it:

```
char atSign;
atSign = '@';
```

The at sign, @, is assigned the decimal Unicode number 64. Behind the scenes, Processing uses this integer to store the at sign in a `char` variable. Thus, we can visualize the result of this program as either

char   `'@'`
     atSign

or

char   `#64`
     atSign

If we add the following call to the `println()` function to our program,

```
char atSign;
atSign = '@';
println(atSign);
```

then this stored character is displayed in the console:

```
@
```

Using values and variables of type char can be a little tricky at times, and they will not be used often in this book. However, it is still very worthwhile to have a basic understanding of the char type so that you know how Processing handles char values when you do need to use them.

Save and close the CharType program.

Multiple Characters: The String Type

Select **File > New** to start a new program. Save this new program as **Strings**.

The char type only allows us to work with a single character. Processing also provides the String type, which allows us to work with a sequence of characters at the same time.

As we have seen, the first letter of the keywords for the int, float, and char types is lowercase. This is because they are examples of what are known as **primitive** types.

In contrast, the first letter of the keyword for the String type name is always *capitalized*. This is because the String type is an example of a more advanced data type known as a **class**. Each specific String value that we work with is sometimes known as a String **object**.*

A value of type String is enclosed in a pair of *double* quotes. All the following are examples of String values:

```
"Processing"
"12345"
"Programming in Processing"
```

A String value can also consist of a single character. Examples are

```
"r"
"9"
"?"
```

---

* We will learn more about classes and objects later in this book.

However, be careful not to confuse a String value that consists of a single character with a char value. Remember, a String value is enclosed in a pair of *double* quotation marks:

    "r"

A value of type char, on the other hand, is enclosed in a pair of *single* quotation marks:

    'r'

A String value can also consist simply of one or more **spaces**:

    " "

A String value known as the **empty string** consists of no characters at all:

    ""

To declare a variable of type String, we simply use a declaration statement of the same form as that which we have used for primitive-type variables:

    String word;

We can visualize the result of this statement as

String ⃞

word

To assign this variable a value of type String, we simply use an assignment statement of the same form as that which we have used for primitive variables, remembering to enclose the characters within a pair of double quotes.

    String word;
    word = "Hello";

Because a String is an *object*, it is stored in the computer's memory in a manner that is different from that used for primitive variables.* However, most of the time, using String variables in Processing is so similar to using primitive-type variables that, in most cases, it is fine for us to visualize a String variable as

word

---

* Specifically, when we assign a String value to a String variable, Processing stores the specific String value in the computer's memory and then stores the address of this String value in the String variable.

We can output the value stored in a `String` variable using the `println()` method in the same manner as we have done before.

Example

```
String word;
word = "Hello";
println(word);
```

When we run the above program, we see the following output to the console:

Hello

Save and close the `Strings` program.

`String` Concatenation

Select **File > New** to create a new program. Save this new program as **Concatenate**.

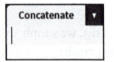

The symbol used to represent addition, +, represents a *different operation* when a value of `String` type is involved. When one or both of the operands is of type `String`, the + operator has the effect of *joining* the operands together into a single `String` value. This operation is known as `String` **concatenation**. For example, the following operation,

```
"blue" + "berry"
```

produces the following result:

```
"blueberry"
```

Remember, if even *one* of the two operands is of type `String`, the + operator is interpreted as concatenation, and the non-`String` operand will be converted into a `String` value in order to complete this operation. Thus, the following expression,

```
"May " + 28
```

evaluates to the following:

```
"May 28"
```

We obtain a similar result even if the non-`String` operand comes first in the operation. Thus, the following expression,

```
28 + " May"
```

evaluates to the following:

```
"28 May"
```

We can verify these results using the `println()` function:

```
println("blue" + "berry");
println("May " + 28);
println(28 + " May");
```

When we run this program, we see the following console output:

blueberry
May 28
28 May

Multiple + operators may be used with concatenation. For example,

```
println("May" + 2 + 8);
```

produces the following output to the console:

May28

Concatenation has the same level of precedence in the order of operations as addition. Thus, when both addition and concatenation appear in the same expression, the operations are simply performed left to right. For example, the following statement,

```
println(2 + 8 + "May");
```

produces the following output to the console:

10May

Why is this the result? When the expression `2 + 8 + "May"` is evaluated left to right, the first operation, `2 + 8`, has two `int` operands. Therefore, the first + operation is interpreted as addition, not concatenation.

Concatenation is also performed if one or both of the operands is a `String` *variable*. For example, given the following variable declaration and initialization,

```
String food;
food = "fruit";
```

consider the following concatenation operations:

```
food + "cake" produces "fruitcake"

"grape" + food produces "grapefruit"

food + 365 produces "fruit365"

365 + food produces "365fruit"
```

We can verify these results using the `println()` function:

```
String food;
food = "fruit";

println(food + "cake");
println("grape" + food);
println(food + 365);
println(365 + food);
```

When we run the above program, we see the following output to the console:

```
fruitcake
grapefruit
fruit365
365fruit
```

Save and close the `Concatenate` program.

Summary

This chapter supplied us with some very important foundations that we will use throughout this book.

- We learned about two *numeric* **types**, `int` and `float`.

- We explored **values** and **variables** of these two types.

- We looked at **arithmetic operators**, **expressions**, and the **order of operations**.

- We learned about how to **assign** values to variables using **assignment statements** containing the **assignment operator** (=).

- We examined the use of *non*numeric **types** of information, individual characters and strings of multiple characters, using specific values and variables of both the `char` and `String` types.

- We learned about how the + operator is interpreted as the **concatenation** operation when one or both of the operands is of type `String`.

Exercises

1) Evaluate each of the following arithmetic expressions. Guess at the answer before using the `print()` or `println()` functions to verify the result of the expression.

```
1 + 2
1 + 2.0
1.0 + 2
6 - 2.0
2.0 + 3
5.0 + 2.0 * 3.0
3 - 1 * 2.0
2 - 3.0 + 1
2.0 + 3 - 2
2 + 1 - 4
2 / 1
2 / 1.0
1.0 / 2
1 / 3
6 / 4.0 - 1
2 + 5 / 2
5 / 3 + 2
3 % 2
2 + 4 * 2
(3 + 2) * 5
5.0 * 2.0
5.0 * 2
5 * 2.0
5 / 11
5 % 11
11 / 5
11 % 5
8 / 3
8 % 3
5 + 4 % 3
```

2) For each of the following, identify whether or not it is a valid variable name in Processing. For any that is not valid, explain why it is invalid.

```
center
column
row
sum
summ
firstRow
firstPixelRow
secondPixel
2ndPixel
second-pixel
second pixel
second_pixel
firstAndOnly
first&only
```

3) Create a declaration statement for a variable named `total` that can store an integer.

4) Create a declaration statement for a variable named `average` that can store a number with decimal places.

5) Create a declaration statement for a variable named `initial` that can store just the first letter of a person's name.

6) Create a declaration statement for a variable named `firstName` that can store a person's first name.

7) Write a program to do the following:

   a) Declare a variable named `angle`.
   b) Assign the value 90 to this variable.
   c) Output the value of this variable to the console.

The output to the console should be exactly the following:

90

8) Write a program that declares a variable to store a person's age as an integer, initializes this variable to 19, and outputs the value of this variable to the console.

9) Write a program that declares a variable to store a person's shoe size, initializes this variable to 10.5, and outputs the value of this variable to the console.

10) Write a program that declares a variable to store only the first letter of a person's first name, initializes this variable to the first letter of your first name, and outputs the value of this variable to the console.

11) Write a program that declares a variable to store a person's first name, initializes this variable to your first name, and outputs the value of this variable to the console.

12) What is the output of the following program? Make a guess before typing in this code and running it.

```
int length;
length = 10;
println(length);
```

13) What is the output of the following program? Make a guess before typing in this code and running it.

```
float length;
length = 100.5;
println(length);
```

14) What is the output of the following program? Make a guess before typing in this code and running it.

```
int amount;
amount = 10 * 3;
println(amount);
```

15) What is the output of the following program? Make a guess before typing in this code and running it.

```
int portion;
portion = 14 / 4;
println(portion);
```

16) What is the output of the following program? Make a guess before typing in this code and running it.

```
float portion;
portion = 14 / 4;
println(portion);
```

17) What is the output of the following program? Make a guess before typing in this code and running it.

```
float portion;
portion = 14 / 4.0;
println(portion);
```

18) What is the output of the following program? Make a guess before typing in this code and running it.

```
int total;
total = 14;
int portion;
portion = total / 4;
println(portion);
```

19) What is the output of the following program? Make a guess before typing in this code and running it.

```
int total;
total = 14;
float portion;
portion = total / 4;
println(portion);
```

20) What is the output of the following program? Make a guess before typing in this code and running it.

```
float total;
total = 14.0;
float portion;
portion = total / 4;
println(portion);
```

21) What is the output of the following program? Make a guess before typing in this code and running it.

```
int inches;
inches = 4 * 3;
println(inches);
```

22) What is the output of the following program? Make a guess before typing in this code and running it.

```
int inches;
inches = 12;
println("Inches: " + inches);
```

23) What is the output of the following program? Make a guess before typing in this code and running it.

```
int feet;
feet = 10;
println("Feet: " + feet);
```

24) What is the output of the following program? Make a guess before typing in this code and running it.

```
int yards;
feet = 2;
int inches;
inches = feet * 12;
println("Inches: " + inches);
```

25) Write a program to do the following:

a) Declare a variable named `first`.
b) Assign 6.5 to `first`.
c) Declare a variable named `second`.
d) Assign 4.0 to `second`.
e) Declare a variable named `total`.
f) Use `first` and `second` to calculate the sum of these two variables and assign the result to `total`.
g) Output "`total:`" to the console, followed by the value of `total`.

The console output should be

```
total: 10.5
```

26) What is the output of the following program? Make a guess before typing in this code and running it.

```
int yards;
yards = 2;
println("Yards: " + yards);

int inches;
inches = yards * 12;
println("Inches: " + inches);
```

27) What is the output of the following program? Make a guess before typing in this code and running it.

```
String first;
String second;
first = "pan";
second = "cake";
println(first + second);
```

28) Write a program to do the following:

    a)   Declare a variable named `greeting` that can hold multiple characters.

    b)   Assign `"hello"` to `greeting`.

    c)   Use `greeting` to produce the following console output:

       `hellohellohello`

29) Write a program to do the following:

    a)   Declare a `String` variable named `first`.

    b)   Assign `first` the value `"watch"`.

    c)   Declare a `String` variable named `second`.

    d)   Assign `second` the value `"your"`.

    e)   Declare a `String` variable named `third`.

    f)   Assign `third` the value `"step"`.

    g)   Use the `first`, `second`, and `third` variables to produce the following three lines of console output:

       `watchyourstep`
       `stepyourwatch`
       `yourwatchstep`

30) Write a program to do the following:

    a)   Declare an `int` variable named `first`.

    b)   Assign `first` the value `2`.

    c)   Declare a `String` variable named `second`.

    d)   Assign `second` the value `"x"`.

    e)   Declare an `int` variable named `third`.

    f)   Assign `third` the value `4`.

    g)   Use the `first`, `second`, and `third` variables to produce the following five lines of console output:

       `2x4`
       `4x2`
       `x42`
       `x24`
       `6x`

CHAPTER 3

More about Using Processing's Built-In Functions

In this chapter, we will continue to learn about Processing's built-in functions. We will see just how useful these particular functions are, and we will also learn more about the nature of functions in general.

More about Console Output: The `print()` and `println()` Functions

Select **File > New** to start a new program. Save this new program as **Console**.

As we have seen, the `println()` function is very useful for displaying a line of information in the console. For example, the following program,

```
println("May");
println(28);
```

produces the following output to the console:

```
May
28
```

Processing also provides the **`print()`** function for displaying output to the console *without* advancing to a new line. For example, the following program,

```
print("May ");
println(28);
```

produces the following output to the console:

```
May 28
```

Notice that after we used the `print()` function to generate output to the console on a single line, we called the `println()` function to display the last item of the current line of output and also to advance to the *next line* in the console:

```
print("May ");
println(28);
```

As a result of this call to the `println()` function, any subsequent output to the console will begin on the line *after* the one containing the date output.

Delete the two lines of code that currently comprise the `Console` program and save this program. We will continue working with this program in the next section.

Displaying Multiple Items to the Console

A single call to the `print()` or `println()` function can also be used to output *multiple* items on a single line in the console. In Processing, there are *two* main techniques for accomplishing this.

The first technique is to separate the items by **commas**. These items will all be displayed on the same line of the console, but a **space** will be *automatically* inserted between each item. For example, the following statement,

```
println("May", 28);
```

produces the following console output:

```
May 28
```

Notice that a space is automatically inserted between the `String` value `"May"` and the `int` value `28`. Similarly, the following statement,

```
println(5, 10);
```

produces the following output to the console:

```
5 10
```

Here, too, notice that a space is automatically inserted between the `int` value `5` and the `int` value `10`.

The second technique for displaying multiple items on a single line in the console is to make use of `String` concatenation. This is the technique that will be used in this book.

For example, the following statement,

```
println("May" + 28);
```

produces this output to the console:

May28

Notice that a space is *not* automatically inserted between the items when String concatenation is used. Thus, any spaces we would like to separate the items must be explicitly included. For example, if we write

```
println("May " + 28);
```

then our two items are now separated by a space when they are output to the console:

May 28

Similarly, if we would like to use concatenation to output two *numeric* values separated by a space, then we can write a statement like the following:

```
println(5 + " " + 28);
```

This statement produces the following output to the console:

5 28

Save and close the Console program.

Graphical Text in Processing

Select **File > New** to start a new program. Save this program as **CanvasText**.

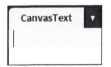

Processing also provides several functions for working with text on the canvas.

The text() Function

The text() function will draw text on the canvas. The basic form of a call to the **text()** function is

```
text(value, column, row);
```

The parameters of the text() function are

    value   The value to be displayed; this can be a value of primitive or `String` type

    column   The pixel column number where the left edge of the text will be displayed on the canvas

    row   The pixel row number where the baseline of the text will be displayed on the canvas

For example, the following program,

```
text("Hi!", width / 2, height / 2);
```

generates the following output to the canvas:

The current **fill color** determines the color of the text that is drawn on the canvas. Because Processing's default fill color is white, any text drawn on the canvas will be white unless we change the fill color using the **fill()** function prior to calling the `text()` function. For example, adding the following statement,

```
fill(0, 0, 0);
text("Hi!", width / 2, height / 2);
```

causes the text to be black when we run the program again:

Save the `CanvasText` program. We will continue working with it in the next section.

The textsize() Function

Return to the `CanvasText` program.

By default, the height of the text drawn by the `text()` function is 12 pixels. To change this height, we can use the `textSize()` function.

The basic form of a call to the `textSize()` function is

```
textSize(pixels);
```

The parameter is

    pixels   The pixel height of the text to be displayed by a subsequent call to the `text()` function. (This will be the pixel height of the tallest characters.)

To illustrate, insert the following call to the textSize() function into our current CanvasText program:

```
fill(0, 0, 0);
textSize(32);
text("Hi!", width/2, height/2);
```

When we run this program, we now see the following text drawn on the canvas:

Save the CanvasText program. We will continue working with it in the next section.

The textAlign() Function

Return to the CanvasText program.

By default, the text drawn by the text() function is *left* aligned relative to the pixel column specified. To change this alignment, we can use the textAlign() function.

The basic form of a call to the textAlign() function is

```
textAlign(align);
```

The parameter is

align   The LEFT, RIGHT, or CENTER constant

To illustrate, insert the following call to the textAlign() function into our current program:

```
fill(0, 0, 0);
textSize(32);
textAlign(CENTER);
text("Hi!", width/2, height/2);
```

When we run this program, we now see the following text drawn on the canvas:

Other functions related to drawing text on the canvas are explained in Processing's Reference.

Save and close the CanvasText program.

Matching the Type of an Argument to the Type of a Parameter

Start a new program by selecting **File > New**. Then, save this new program with the name **Functions**.

In most of the calls to Processing's drawing functions that we have made so far, we have used *integers* as the arguments. For example, in Chapter 2, we made calls to the `size()` function like the following:

```
size(150, 200);
```

Actually, the `size()` function is one of the few Processing drawing functions that *requires* us to supply an `int` value for each of its parameters. For example, if we write the following,

```
size(150.0, 200.0);
```

then Processing underlines the beginning of this statement:

```
size(150.0, 200.0);
```

Also, an error message appears in the Message Area:

The function "size()" expects parameters like: "size(int, int)"

Processing is alerting us to the fact that the `size()` function has two *parameters*, both of which are `int` type. Thus, the two *arguments* that we supply to meet these parameter requirements of this function must both be `int` type as well. In other words, every time we call the `size()` function, we need to specify two `int` values. In this case, we received an error because we instead specified two `float` values as the two arguments. As we have learned, a `float` value cannot be used when an `int` value is expected.

How are we to determine the *types* of the parameters required by a particular Processing function so that we can know what types of arguments we need to use when we call that function? Rest assured, from this point forward in this book, whenever a function is used for the first time, the types of any parameters it has will always be identified!

However, another option is to check the **Reference** that is built into Processing. To view the Reference, select **Help > Reference**. In Processing's Reference, we can look up the **parameter list** for any of Processing's functions. For example, if we look up the `size()` function in Processing's Reference, we see the syntax (basic form) of this function:

Syntax   `size(w, h)`

This entry in the Reference also tells us not only *how many* parameters are required but also the *type* of value that is expected for each of the parameters of the size() function. A brief explanation of each parameter is also provided:

Parameters: w   **int:** width of the display window in units of pixels
                    h   **int:** height of the display window in units of pixels

Here we see that the size() function does indeed have two parameters of type int. Thus, we must specify two arguments of type int whenever we call this function.

Similarly, if we look up the point() function in the Processing Reference, we see that the syntax (basic form) of this function is:

Syntax   point(x, y)

Once again, the Reference entry also tells us what *type* of value is expected for each of the parameters of the point() function, along with a description of each parameter:

Parameters: x   **float:** x-coordinate of the point
                    y   **float:** y-coordinate of the point

Here, we see that the point() function has two parameters of type float. Thus, we must specify two arguments of type float whenever we call this function.

As it turns out, unlike the size() function, almost all of Processing's other drawing functions have parameters that are float type. In other words, these functions expect us to specify float arguments whenever we call them.

For example, because the point() function expects arguments of type float, we can write

```
point(50.2, 75.5);
```

and we will not receive an error message. However, as we have learned, Processing will use only *whole* pixels when it draws on the canvas, so it will essentially ignore any decimal places in these float arguments. Thus, the above call to the point() function will essentially become

```
point(50.0, 75.0);
```

In most of our calls to the drawing functions prior to this chapter, we used int arguments, even though all these functions (except for the size() function) were expecting float arguments. This didn't cause an error because, as we have learned,

an int value can be supplied wherever a float value is expected.

Whenever we supply an int argument for a float parameter, that int value will be automatically converted by Processing to a float value. It's as if Processing simply appends a .0 on the end of the int value. Thus, when we make a function call such as

```
point(50, 75);
```

this function call is essentially converted by Processing into the following:

```
point(50.0, 75.0);
```

We will learn more about matching an argument's type to a parameter's type when we define our own functions later in this book.

Two Kinds of Functions

We have already used quite a variety of Processing's built-in functions up to this point, most of which have pertained to drawing on the canvas. However, all the functions we have used so far have been examples of what are known as void functions. There are actually *two* main categories of functions in Processing:

1) void functions

2) (Non-void) functions

Let's learn more about both of these kinds of functions.

void Functions

A **void function** is like a little machine that has a predefined action. We simply call the function, supplying any arguments that it needs, and the function will perform its predefined action for us.

void function

The keyword void is somewhat cryptic, but don't worry about it for now. We'll look more closely at this term in the next section.

To call a void function, we use the *name* of the function along with a pair of parentheses enclosing any *arguments* that are required, followed by a semicolon:

```
function(arguments);
```

For example, whenever we call the size() function, we simply write something like

```
size(150, 100);
```

Whenever a `void` function is called, it performs its predefined action. Once that predefined action is finished, the call to the function is complete. Thus, when the `size()` function is called, it simply performs its predefined action: it generates a canvas that has pixel dimensions equal to whatever arguments we supply when we call this function. Once this canvas is generated, the call to the `size()` function is complete. Processing then moves on to the next statement in the program. We can visualize the action of calling this void function as

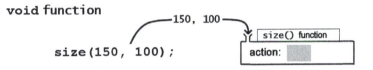

All of Processing's *drawing* functions are `void` functions. However, there are quite a few other `void` functions in Processing as well. For example, the `println()` function is also a `void` function. Remember: The general form of a `void` function call is

function(arguments);

Thus, the `println()` function can be called simply by giving the name of the function along with a pair of parentheses enclosing any necessary arguments, followed by a semicolon:

```
println("Hello");
```

When the `println()` function is called, it simply performs its predefined action by displaying in the console whatever arguments we have supplied to this function and then advancing to the next line of the console:

Hello

After this action of outputting to the console, the call to the `println()` function is complete. Processing then moves on to the next statement in the program. We can visualize the action of calling this `void` function as

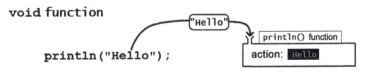

In summary, the main characteristics of using a **void function** are the following:

- When a `void` function is called, it simply performs whatever action it is predefined to perform.

- A statement containing a call to a `void` function consists of the name of the function, along with a pair of parentheses containing any necessary arguments, followed by a semicolon.

  `function(arguments);`

  Example

  `println("Hello!");`

Functions That Return a Value

Not all of Processing's functions are `void` functions. Processing also has non-`void` functions. These are known simply as **functions**. Such a function is similar to a `void` function in that it is like a little machine that performs a predefined action when we call it. However, this kind of function also performs an *extra step*: it **returns a resulting value** to us.

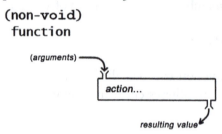

For example, the basic form of the `sqrt()` function is

`sqrt(number)`

The parameter is

number   Any nonnegative value of type `float`. (Remember: If an `int` value is specified, Processing will automatically convert this integer into a `float` value.)

When we call the `sqrt()` function, it performs its predefined action: it calculates the square root of the `float` argument that we supply to it. However, this function wouldn't be very useful if did not also report to us what that calculated square root is. Thus, unlike a `void` function, the `sqrt()` function performs the extra step of *returning* to us a resulting item of information. In the case of the `sqrt()` function, this resulting item of information returned is the square root that the function has calculated for us.

Examples

```
sqrt(9.0)    returns 3.0
sqrt(9)      returns 3.0
sqrt(6.25)   returns 2.5
```

We can visualize the action of calling this function as

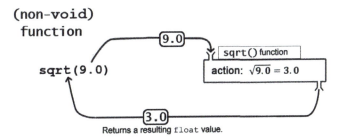

Returns a resulting float value.

The value that is returned by a non-void function is always a predefined *type* of value. For example, the value returned by the sqrt() function is always a float value. For this reason, we say that the sqrt() function's **return type** is float.

Because a non-void function performs an extra step of returning a resulting value, we also need to perform an extra step when we *call* such a function so that we do not *lose* the resulting value that is returned to us. To illustrate, suppose that we mistakenly call the sqrt() function using the technique that we use for calling a void function:

```
sqrt(9.0);
```

This call does cause the sqrt() function to both calculate the square root of 9.0 and return the resulting float value, 3.0, to us.

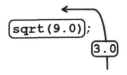

However, because we have done nothing to make use of this returned result, it is lost. It's now too late for us to make any further use of the square root value that was returned to us.

For this reason, whenever we call a (non-void) function that returns a value, we immediately need to *make use of* **the returned value**. A very good way do this is by assigning the returned value to a variable. Thus, the general form we will typically use when calling a non-void function is

```
variable = function(arguments);
```

This way, we do not lose the resulting value that is returned to us by the function, because we are now storing this returned value in a variable. We can then use this variable to make further use of the returned value as many times as we like.

So, let's declare a variable to hold the resulting square root that is returned to us by the sqrt() function. Because the return type of the sqrt() function is float, we know that the square root returned by this function will always be a float value. Thus, any variable we use to hold this returned square root needs to be a `float` variable. For example, we can declare the following variable:

```
float root;
```

We can visualize the result of this variable declaration as

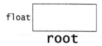

We can now use this `float` variable, `root`, to store the `float` value returned by the `sqrt()` function simply by creating an assignment statement like the following:

```
float root;
root = sqrt(9.0);
```

Let's step through this assignment statement. As always, the *right*-hand side of an assignment statement is evaluated *first*. In this particular case, the right-hand side contains a call to a function that will return a value:

$$\text{root} = \boxed{\text{sqrt(9.0)}};$$

Thus, when the right-hand side of this assignment statement is evaluated, the **sqrt()** function is called using the argument supplied, 9.0. This function then performs its predefined action of calculating the square root of 9.0 and then performs the additional step of returning to us this resulting `float` value, 3.0. We can visualize this returning of a value as

$$\text{root} = \boxed{\text{sqrt(9.0)}};$$

The evaluation of the right-hand side of the assignment statement is now complete. Thus, the assignment essentially becomes

$$\text{root} = \boxed{3.0};$$

Thus, the returned `float` value is now stored in the `root` variable, which is also of type `float`. We can visualize this assignment operation as

$$\underset{\text{(float)}}{\text{root}} = \underset{\text{(float)}}{\boxed{3.0}};$$

And we can visualize the result of this assignment statement as

float | **3.0**
root

Because we stored the returned value in a variable, we are now able to make additional use of the returned value by means of this variable. For example, we can display the resulting square root in the console simply by adding the following statement:

```
float root;
root = sqrt(9.0);
println("The square root is: " + root);
```

When we run this program, we now see the following output to the console:

```
The square root is: 3.0
```

In summary, the main characteristics of using a (non-void) **function** are the following:

- When a non-void function is called, it performs the action it is designed to do, but it also **returns a resulting value**.

- The value returned by a non-void function is always of a specific *type*. This is known as the function's **return type**.

- A statement containing a call to a non-void function should *make use* of the value that is returned, typically by assigning this returned value to a *variable*:

```
variable = function(arguments);
```

Example

```
float root;
root = sqrt(9.0);
```

You may save and close the Functions program.

Determining a Function's Return Type Using Processing's Reference

How do we determine the *type* of value that is *returned* to us by a particular function? From this point forward in this book, the return type of a function will always be identified. However, the Reference built into the Processing IDE contains this information as well. For example, you may have noticed in the Processing Reference that the entries for both the size() and point() functions include the following:

Returns   void

This particular item of information appears in the Reference entries for both size() and point() because both of these functions are examples of void functions.

In contrast, if we open the Processing Reference and look up the sqrt() function, the entry includes an item of information telling us that this function returns a float value:

Returns: **float**

Thus, we know that the sqrt() function is *not* a void function. Rather, the sqrt() function's return type is float. In other words, the sqrt() function always returns a resulting value of type float to us whenever we call this function.

Example: Calculating a Hypotenuse with the sqrt() Function

A right triangle consists of three sides: two *legs* on either side of the right angle and the *hypotenuse*, which is opposite the right angle.

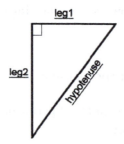

Let's write a program to draw such a triangle. Start a new program by selecting **File > New**. Save this program with the name **Triangle**.

Recall that the triangle() function will render a triangle on the canvas if we give it the column number and row number for each of the three pixel locations where we want the points (vertexes) of the triangle to be. For example, suppose we wish to draw the following triangle on a 300-pixel by 300-pixel canvas:

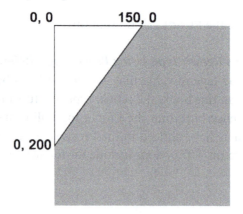

We can draw this triangle by writing the following:

```
size(300, 300);
triangle(0, 0, 150, 0, 0, 200);
```

Or, we can draw the same triangle by using two variables to store the lengths of the two legs of the triangle and then use these two variables in the call to the `triangle()` function:

```
size(300, 300);

int leg1;
leg1 = 150;

int leg2;
leg2 = 200;
triangle(0, 0, leg1, 0, 0, leg2);
```

When we run this version of the program, we see the same triangle rendered on the canvas as before:

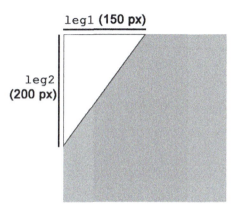

The above program includes variables that store the lengths of the two legs of the right triangle. Let's calculate the length of the hypotenuse as well. According to the Pythagorean theorem, the following is true:

$$\text{hypotenuse}^2 = \text{leg1}^2 + \text{leg2}^2$$

Therefore, the following is also true:

$$\text{hypotenuse} = \sqrt{\text{leg1}^2 + \text{leg2}^2}$$

Thus, we can use Processing's `sqrt()` function to calculate the length of the hypotenuse.

Whenever we use a function that returns a value, we will declare a variable to hold this returned value. We know that the return type of the `sqrt()` function is `float`, so we need to declare a `float` variable to hold the resulting value returned to us by this function:

```
float hypotenuse;
```

Once we have declared this variable, we can add a statement to calculate the length of the hypotenuse using the sqrt() function and store this length in the hypotenuse variable:

```
float hypotenuse;
hypotenuse = sqrt((leg1 * leg1) + (leg2 * leg2));
```

Finally, let's add a statement to display the calculated hypotenuse length to the console:

```
float hypotenuse;
hypotenuse = sqrt((leg1 * leg1) + (leg2 * leg2));
println("Hypotenuse: " + hypotenuse);
```

Our modified program is now

```
size(300, 300);

int leg1;
leg1 = 150;

int leg2;
leg2 = 200;

triangle(0, 0, leg1, 0, 0, leg2);

float hypotenuse;
hypotenuse = sqrt((leg1 * leg1) + (leg2 * leg2));
println("Hypotenuse: " + hypotenuse);
```

When we run the program, we see the same triangle as before drawn on the canvas, but we now also see the length of the hypotenuse displayed in the console:

```
Hypotenuse: 250.0
```

Save this Triangle program. We will return to this example later in this chapter.

The pow() Function

Processing's pow() function can be used to raise a value to a power. It is another example of a function that returns a value. The basic form of a call to the pow() function is

```
pow(base, exponent)
```

The parameters are

base   The number to be raised to a power. Type: **float**.

exponent   The exponent to which the number is to be raised. Type: **float**.

The type of the value returned by pow() is **float**.
For example, to calculate 10^2—that is, to raise 10 to the power 2—we can write

```
pow(10.0, 2.0)
```

The *base* and *exponent* arguments of the pow() function are expected to be of type float. However, remember that an **int** value may be used as an argument whenever a **float** value is expected, because Processing will automatically convert that int value to a float value. Thus, we may also write

```
pow(10, 2)
```

This will not cause an error, because Processing automatically converts the int values in this function call, 10 and 2, to the float values 10.0 and 2.0.

Calculating the Area of a Square Using the pow() Function
Start a new Processing program by selecting **File > New**. Save this program as **Square**.

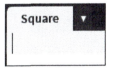

Enter the following lines of code:

```
size(200, 200);

int side;
side = 150;

rect(0, 0, side, side);
println("Side length of square: " + side);
```

When we run this program, it draws a square with a side length of 150 pixels on a canvas that is 200 pixels by 200 pixels:

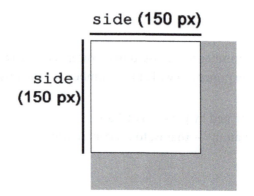

It also displays the side length of the square to the console:

```
Side length of square: 150
```

We can modify this program so that it also uses a variable to calculate and display the area of this square simply by adding the following:

```
size(200, 200);

int side;
side = 150;

rect(0, 0, side, side);
println("Side length of square: " + side);

int area;
area = side * side;
println("Area of square: " + area);
```

When we run this modified version of our program, the square that is drawn on the canvas is the same size as before:

However, we now see that the area of the square is also output to the console:

```
Side length of square: 150
Area of square: 22500
```

Let's instead calculate the area of the square by making the following change:

```
size(200, 200);

int side;
side = 150;

rect(0, 0, side, side);
println("Side length of square: " + side);

int area;
area = pow(side, 2);
println("Area of square: " + area);
```

Unfortunately, as soon as we make this change, our call to the pow() function is now underlined,

```
area = pow(side, 2);
```

and we see the following error message in the Message Area:

Type mismatch, "float" does not match with "int"

(As always, if you do not see the error message in the Message Area, then click on the under-lined portion of the code, and it will appear.) Processing is alerting us to the fact that the pow() function returns a resulting value of type float. However, we have attempted to assign this returned float value to the area variable, which is currently declared to be of type int. Thus, we have a type mismatch in our assignment statement, one that we can visualize as:

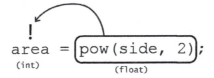

We can eliminate this type mismatch simply by declaring the area variable to be of type float:

```
size(200, 200);

int side;
side = 150;

rect(0, 0, side, side);
println("Side length of square: " + side);

float area;
area = pow(side, 2);
println("Area of square: " + area);
```

After making this change, we no longer see an error message. When we run this program, notice that the output to the console now includes a float value for the area of the square:

```
Side length of square: 150
Area of square: 22500.0
```

Save this Square program. We will return to it later in this chapter.

Calculating the Hypotenuse with the pow() *Function*

Let's return to our Triangle program. In it, we used the following statement to calculate the hypotenuse:

```
hypotenuse = sqrt((leg1 * leg1) + (leg2 * leg2));
```

However, let's now modify this statement to make use of the pow() function:

```
size(300, 300);

int leg1;
leg1 = 150;

int leg2;
leg2 = 200;

triangle(0, 0, leg1, 0, 0, leg2);

float hypotenuse;
hypotenuse = sqrt( pow(leg1, 2) + pow(leg2, 2) );
println("Hypotenuse: " + hypotenuse);
```

This is another example of **nesting** a call to one function inside a call to another function. Nesting a function call is another way that we can make immediate use of the value that is returned by that function so that it is not lost. When we rerun the program, we obtain the same output to the canvas as before:

We also see the same output to the console as before:

```
Hypotenuse: 250.0
```

Save the `Triangle` program and close it.

The `random()` Function

The `random()` function enables us to pick a random number from a range of values. This function is often very useful when creating a simulation, a game, or an animation. The basic form of a call to the **random()** function is

```
random(minimum, maximum)
```

The parameters are

minimum	The minimum number for the range. Type: **float** . If this argument is *omitted*, then the minimum is assumed to be *zero*.
maximum	The maximum number for the range. Type: **float**. The random number chosen will be *greater than or equal to* **minimum** but strictly *less than* **maximum**.

The type of value returned is **float**.

Let's clarify the use of the `random()` function with some examples. The call to the `random()` function

```
random(1, 10)
```

will generate and return to us a random `float` value that is *greater than or equal* to 1 but strictly *less than* 10. In other words, the random value chosen for us will be from (and including) 1 up to (but *not* including) 10. Similarly, the function call

```
random(0, 20)
```

will generate and return to us a random `float` value that is *greater than or equal* to 0 but strictly *less than* 20. In other words, the random `float` value chosen for us will be somewhere in the range from (and including) 0 up to (but *not* including) 20.

As noted above, when the minimum value is zero, we are allowed to omit the first argument. Thus, the previous function call can also be written as simply

```
random(20)
```

Open a new program in Processing by selecting **File > New**. Name this new program **Random**.

Enter the lines of code shown below to create a program that uses the random() function to generate a random float value from 1 up to (but not including) 10 and display it in the console:

```
float number;
number = random(1, 10);
println("Number chosen: " + number );
```

When this program is run, its output is similar to the following:

Number chosen: 5.5593557

(The actual float value that you see will most likely be different from the one shown here.)

Save the Random program. We will continue working with it in the next section.

The round() Function

The round() function rounds a float value to the nearest integer. The basic form of a call to this function is

round(*number*)

The parameter is

number   The number to be rounded. Type: **float**.

The type of value returned is **int**.

If the decimal portion of the float value supplied as an argument to this function is *less than* .5, then this float value is rounded *down* to the *previous* integer.

Examples

```
round(6.0)   returns the int value 6
round(6.3)   returns the int value 6
round(6.49) returns the int value 6
```

On the other hand, if the decimal portion of the float value supplied as an argument to this function is *greater than or equal to* .5, then this float value is rounded *up* to the *next* integer.

Examples

```
round(6.5) returns the int value 7
round(6.7) returns the int value 7
```

Using the `round()` *Function*

Return to the **Random** program.

The following program illustrates the use of both the `random()` function and the `round()` function:

```
float number;
number = random(1, 10);
println("Number chosen: " + number);

int rounded;
rounded = round(number);
println("Rounded: " + rounded);
```

Each time we run this program, a random unrounded number is selected and displayed in the console along with the rounded version of this same number:

```
Number chosen: 7.3311152
Rounded: 7
```

Notice that the above unrounded random `float` value was rounded *down* to the previous integer. In the following sample run, the random `float` value was rounded *up* to the next integer:

```
Number chosen: 8.706958
Rounded: 9
```

Save and close the `Random` program.

More Conversion Functions: `int()` *and* `float()`

Processing provides a number of other functions that will convert a value of one type to a value of another type. Here, we will look at two such functions: `int()` and `float()`.

The `int()` Function

The basic form of a call to the **int()** function is

> **int(*value*)**

The parameter is

> **value**   The value that is to be converted to an `int` value.
>
>   Type: A primitive type or a `String` type that is comprised entirely of numerals.

The type of value returned is `int`.

The most common use of the `int()` function is to convert a `float` value to an `int` value. When used for this purpose, the `int()` function essentially truncates (cuts off) the fractional part of a decimal number, producing an integer. In other words, any digits to the right of the decimal point are essentially removed, and an `int` value results.

Examples

```
int(6.2)     returns the int value 6
int(6.7)     returns the int value 6
int(6.0)     returns the int value 6
int(6.99)    returns the int value 6
```

The `int()` function can be used not only to convert values of type `float` but also to convert other types of values into `int` values. For example, we can use the `int()` function to convert a `String` of numerals into a corresponding `int` value. The `String` value `"365"` resembles an integer, but it is not an `int` value. Thus, we would not be able to use `"365"` in an arithmetic operation. However, the following function call,

```
int("365")
```

returns the `int` value

```
365
```

We would then be able to use `365` in an arithmetic operation. This ability to convert a `String` value to the `int` value it resembles will be very useful to us later in this chapter when we learn how to *input* `int` values.

`int()` *and* `random()` *Example: Choosing a Random Integer to Simulate Rolling a Die*
Create a new program by selecting **File > New**. Name this new program **Roll**.

Suppose that we wish to simulate the roll of a die by randomly generating an integer from 1 to 6. The following function call,

```
random(1, 7)
```

will return a `float` value from 1 up to (but not including) 7. This means that the lowest possible random value is 1.0, and the highest possible random value is 6.9999999. Thus, we can begin our program with

```
float randomValue;
randomValue = random(1, 7);
```

For example, if we wish to have an int value from 1 to 6 to display in the console, we can use the int() function for this purpose. For this we can write

```
float randomValue;
randomValue = random(1, 7);

int roll;
roll = int(randomValue);

println("Roll: " + roll);
```

Each time that we run this program, a random integer from 1 to 6 is output to the console:

Roll: 3

We can create a shorter version of this program if we *nest* the call to the random() function inside the call to the int() function. In this case, our program becomes

```
int roll;
roll = int( random(1, 7) );

println("Roll: " + roll);
```

When we run this modified version of the program, we still see a random integer from 1 to 6 displayed in the console:

Roll: 5

This nested technique for generating a random integer will be used often in this book.
    Save this Roll program. We will return to this example later in this book.

int() *and* random() *Example: Random Side Length for a Square*

Let's return to our Square program from earlier in this chapter. It currently consists of the following:

```
size(200, 200);

int side;
side = 150;

rect(0, 0, side, side);
println("Side length of square: " + side);

float area;
area = pow(side, 2);
println("Area of square: " + area);
```

When we run this program, we see the same square drawn on the canvas as before:

We also see the same console output as before:

```
Side length: 150
Area of square: 22500.0
```

Next, let's modify this program so that it chooses a side length at random. Suppose we would like this side length to be from 50 up to (but not including) 150 pixels. We can start by modifying our assignment statement to the following:

```
size(200, 200);

int side;
side = random(50, 150);

rect(0, 0, side, side);
println("Side length of square: " + side);

float area;
area = pow(side, 2);
println("Area of square: " + area);
```

However, as soon as we make this change, we see that our call to the random() function is underlined,

```
side = random(50, 150);
```

and we see the following error message displayed in the Message Area:

Type mismatch, "float" does not match with "int"

Processing is alerting us to the fact that the random() function has a return type of float. In other words, the random number generated and returned to us by a call to the random() function is always a float value. However, we cannot assign a float value to the int variable side. We can visualize the type mismatch in this assignment operation as

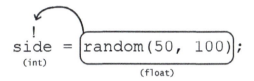

We could resolve this error by changing the type of side to float, because the rect() function in fact expects arguments that are float values. However, Processing will always draw the rectangle using whole pixels, essentially ignoring the decimal places of any float argument that we supply to the rect() function. Thus, it makes more sense for us to choose a random int value for our int variable side, especially if we want our calculated area to be accurate.

We can obtain a random int value if we use the technique we learned in the previous section of nesting our call to the random() function inside a call to the int() function:

```
size(200, 200);

int side;
side = int( random(50, 150) );

rect(0, 0, side, side);
println("Side length of square: " + side);

float area;
area = pow(side, 2);
println("Area of square: " + area);
```

Now, each time we run our program, we see a square with a randomly chosen side length drawn on the canvas:

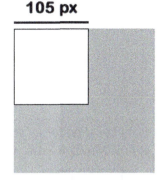

We also see that the side length of the square in pixels is displayed in the console as an integer:

```
Side length: 105
Area of square: 11025.0
```

The area of the square displayed will always be a float value that ends in .0, because this area is now calculated using an int value for side. To have the area be displayed as an integer as well, we can simply change the type of the area variable to int and nest our call to the pow() function inside a call to the int() function:

```
size(200, 200);

int side;
side = int( random(50, 150) );

rect(0, 0, side, side);
println("Side length of square: " + side);

int area;
area = int( pow(side, 2) );
println("Area of square: " + area);
```

Now, when we run our program, a square of a randomly chosen side is drawn on the canvas:

The side length and area are again displayed in the console, but the area is now an integer, too:

```
Side length of square: 72
Area of square: 5184
```

Save the Square program and close it.

Example: Randomly Choosing a Pixel Location for a Circle

Start a new program by selecting **File > New**. Save this program as **Circle**.

Let's work through an example that involves drawing a circle at a random location on the canvas. We'll start with a canvas that is 150 pixels by 200 pixels:

```
size(150, 200);
```

This statement produces the following canvas when we run our program:

On this canvas, we're going to draw a circle with its center at a randomly chosen pixel location on the canvas. Let's declare an `int` variable named `column` to store the randomly chosen pixel column.

```
size(150, 200);

int column;
```

Next, let's assign to this `column` variable a randomly chosen pixel column from 0 up to the width of the canvas. For this, we can write

```
size(150, 200);

int column;
column = int( random(0, width) );
```

or, more simply,

```
size(150, 200);

int column;
column = int( random(width) );
```

Let's look more closely at this last statement. The function call

```
random(width)
```

returns a random `float` value from 0 .0 up to (but not including) 150 .0. This returned value is then converted to an `int` value by the call to the `int()` function:

```
int( random(width) )
```

Thus, the result is an `int` value from 0 to 149. This is a good match to our canvas. The canvas is 150 pixels wide.

And, because the first pixel column of the canvas is column 0, this means that the last pixel column of the canvas is column 149.

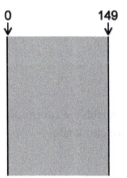

This randomly chosen int value is then assigned to the column variable:

```
column = int( random(width) );
```

Our program is now

```
size(150, 200);

int column;
column = int( random(width) );
```

Next, let's declare an int variable named row to store the randomly chosen pixel row and assign to this variable a randomly chosen pixel column from 0 up to the height of the canvas:

```
size(150, 200);

int column;
column = int( random(width) );

int row;
row = int( random(height) );
```

The height of our canvas is 200 pixels.

200 px

Thus, this last statement,

```
row = int( random(height) );
```

has the effect of assigning to the `row` variable a random `int` value from `0` to `199`. Once again, this is a good fit to our canvas, because it consists of 200 rows of pixels that are numbered from 0 to 199:

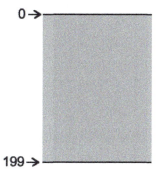

0→

199→

Let's also display the values chosen for `column` and `row` to the console:

```
println(column + ", " + row);
```

We can now use these two variables to draw a circle with diameter of 20 pixels at the randomly chosen canvas location:

```
ellipse(column, row, 20, 20);
```

Our program is now

```
size(150, 200);

int column;
column = int( random(width) );

int row;
```

```
row = int( random(height) );

println(column + ", " + row);

ellipse(column, row, 20, 20);
```

When we run this program, we see the randomly chosen pixel column and pixel row location displayed in the console,

71, 106

and a circle drawn at this same pixel column and pixel row on the canvas:

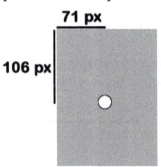

Save the Circle program. We will continue working with it in the next section.

Random Color
Return to the Circle program. If we would also like the fill color of the circle to be chosen at random, we can use the random() function for this as well.

The fill() function expects us to supply three arguments:

fill(*red, green, blue*);

Typically, we have used int values for these three arguments. However, as we have learned, most of Processing's drawing functions actually expect float arguments, and this includes the fill() function. Thus, a function call such as

```
fill(255, 255, 255);
```

will essentially be converted by Processing to

```
fill(255.0, 255.0, 255.0);
```

Also, as with other drawing functions, any decimal places in float arguments to the fill() function will be essentially ignored by Processing. Thus, a function call such as

```
fill(255.123, 255.456, 255.789);
```

will essentially be treated by Processing as

```
fill(255.0, 255.0, 255.0);
```

To choose an appropriate argument at random for the fill() function, we can write

```
random(256)
```

This function call will return a float value from 0.0 up to but not including 256.0, which fits perfectly to the range of values from 0 to 255 that the fill() function expects for each of its arguments. Thus, to pick a color at random, we can nest three such calls to the random() function to specify each of the three RGB arguments for the fill() function:

```
fill( random(256), random(256), random(256) );
```

If we add this statement to our program,

```
size(150, 200);

int column;

column = int( random(width) );

int row;
row = int( random(height) );

println(column + ", " + row);

fill( random(256), random(256), random(256) );
ellipse(column, row, 20, 20);
```

then, when we run our program, we still see a random pixel location chosen for the circle and displayed in the console:

```
45, 147
```

But now, the fill color of the circle is chosen at random:

Save this Circle program. We will make use of it again in a later chapter.

The `float()` Function

Processing also provides the `float()` function for converting values of another primitive type into a value of `float` type. The basic form of a call to the `float()` function is

```
float(value)
```

The parameter is

    value   The value that is to be converted to a `float` value.

    Type: a primitive value or a `String` that resembles a number with decimal places.

The type of value returned is **float**.

    The most common use of the **float()** function is to convert an `int` value to a `float` value. We can think of this as Processing simply appending `.0` onto the end of the integer digits.

Examples

```
float(2)     returns 2.0

float(100)  returns 100.0

float(0)     returns 0.0
```

Using the `float()` function to convert an `int` value to a `float` value provides a way to avoid integer division. Examples:

```
3 / 2              evaluates to 1

float(3) / 2  evaluates to 1.5
```

The `float()` function can also be used to convert values that are of types other than `int` to an equivalent `float` value. For example, `"98.6"` resembles a `float` value, but it is actually a `String` value. Thus, we cannot use it in an arithmetic operation. However, we can use the `float()` function to convert a `String` of numerals into a corresponding `float` value:

```
float("98.6")
```

returns the `float` value 98.6, which can be used in an arithmetic operation.

Like the int() function, the ability to use the float() function to convert a String value into the float value it resembles will prove to be very useful later in this chapter when we learn how to *input* float values.

Using the* float() *Function: Dividing Two* int *Variables Accurately
Start a new program by selecting **File > New**. Save this program as Ratio.

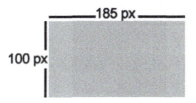

As was mentioned previously, the float function can be particularly useful when we are trying to avoid difficulties that may be produced by integer division. For example, the ratio of the width to the height of an image is sometimes known as its **aspect ratio**. Many movie screens in the United States have an aspect ratio of 1.85:1, meaning that the width of the screen is 1.85 times the height of the screen. We can create a canvas with this aspect ratio by writing

```
size(185, 100);
```

When we run this program, we see the following canvas rendered:

```
————185 px————
|
100 px
|
```

Let's calculate the aspect ratio of the canvas. First, we'll use Processing's width and height variables to display the canvas size.

```
size(185, 100);

println("width: " + width);
println("height: " + height);
```

We'll also need a float variable to store the aspect ratio:

```
size(185, 100);

println("width: " + width);
println("height: " + height);

float ratio;
```

Next, to calculate, store, and display the aspect ratio, let's try the following:

```
size(185, 100);

println("width: " + width);
println("height: " + height);

float ratio;
ratio = width / height;
println("aspect ratio: " + ratio);
```

When we run this program, we see the same canvas displayed as before, but we can see from the console output that our calculation of the aspect ratio is incorrect:

```
width: 185
height: 100
aspect ratio: 1.0
```

Our aspect ratio should be 1.85, not 1.0. What went wrong?

Let's look at our assignment statement more closely. As always, the expression on the right-hand side of the assignment operator (=) is evaluated first:

$$\text{ratio} = \boxed{\text{width / height}};\quad \text{int}\ \boxed{\textbf{185}}\quad \text{int}\ \boxed{\textbf{100}}$$
$$\qquad\qquad\qquad\qquad\qquad \textbf{width}\qquad\quad\ \textbf{height}$$

As we know, both width and height are variables of type int. Thus, the division operation in this expression is integer division:

$$\begin{array}{cc} \text{(int)} & \text{(int)} \\ \textbf{185} & /\ \textbf{100} \end{array}$$
$$\text{ratio} = \boxed{\text{width / height}};$$

Remember, with integer division, any decimal portion of the quotient that would normally result from a division operation is lost. Thus, in integer division, 185 ÷ 100 = 1, not 1.85. This is why, our expression on the right-hand side evaluates to 1:

$$\textbf{1}$$
$$\text{ratio} = \boxed{\textbf{185}\ /\ \textbf{100}};$$

This means that our assignment statement contains a type mismatch, because we are assigning an int value 1 to a float variable, ratio.

$$\text{ratio} = \boxed{\textbf{1}};$$
$$\text{(float)}\qquad \text{(int)}$$

However, as we know, if an int value is supplied where a float value is expected, Processing automatically converts the int value to a float value, essentially appending

".0" onto the end of the int value. Here, this automatic conversion resolves the type mismatch, and the float value 1.0 is stored in the float variable ratio. We can visualize this automatic conversion and assignment operation as

$$
\begin{array}{c}
\text{(float)} \\
\overset{\displaystyle 1.0}{\overbrace{\qquad}} \\
\texttt{ratio} = \boxed{1}\ ; \\
\text{(float)} \quad \text{(int)}
\end{array}
$$

And we can visualize the result of this assignment statement as

$$
\texttt{float}\ \boxed{\textbf{1.0}}
$$
$$
\texttt{ratio}
$$

This explains the result we see in the console: 1.0.

How should we modify our program so that it produces the correct aspect ratio, 1.85? Remember, the float() function can be used to convert an int value to a float value. Thus, let's insert the following calls to the float() function:

```
size(185, 100);

println("width: " + width);
println("height: " + height);

float ratio;
ratio = float(width) / float(height);
println("aspect ratio: " + ratio);
```

When copies of the values stored in the width and height variables are retrieved, the expression on the right-hand side of the assignment operator (=) now becomes essentially

$$
\texttt{ratio} = \boxed{\textbf{float(185) / float(100)}}
$$

Thus, the division operation in our assignment statement becomes float division instead of integer division.

$$
\texttt{ratio} = \boxed{\textbf{185.0 / 100.0}}
$$

As a result, the expression on the right evaluates to 1.85, a float value. Thus, our assignment statement essentially becomes

$$
\texttt{ratio} = \boxed{\textbf{1.85}};
$$

Because both the variable on the left-hand side of the assignment operator and the value on the right-hand side of the assignment operator are of type float, this assignment operation can be completed. We can visualize this assignment operation as

$$
\begin{array}{c}
\texttt{ratio} = \boxed{\textbf{1.85}}; \\
\text{(float)} \qquad \text{(float)}
\end{array}
$$

And we can visualize the result of this assignment statement as

float │ **1.85**
ratio

Thus, when we run our program, we see the same canvas generated as before, but we now see the following output to the console which includes the correct aspect ratio:

```
width: 185
height: 100
aspect ratio: 1.85
```

As we have seen here, the `float()` function is useful for producing `float` division when both of the operands are `int` values. In fact, because `float` division will result when even *one* of the operands is a `float` value, we could have made just one call to the `float()` function and still achieved our desired result. For example, we could have written our assignment statement as simply

```
ratio = float(width) / height;
```

The division operation then becomes

```
185.0 / 100
```

which still evaluates to `1.85`, our desired result.
Save and close the `Ratio` program.

Converting to Nonnumeric Types: The `char()` and `str()` Functions

Processing provides two other conversion functions for converting from one type to another: the `char()` function and the `str()` function.

Select **File > New** to start a new program. Save this program as **NonNumeric**.

NonNumeric ▼

The `char()` Function

Processing's `char()` function can be used to return the `char` value that corresponds to a specified decimal Unicode number. The basic form of a call to the `char()` function is

```
char(unicode)
```

The parameter is

unicode   The Unicode number for the desired character

The type of value returned is **char** .

One use of the **char()** function is to display a character on the canvas or console that is not on the keyboard. For example, the division character (÷) is not on a standard computer keyboard. However, because the Unicode number (in decimal) of this character is 247, the following program,

```
println( char(247) );
```

produces the following output to the console:

```
÷
```

Delete this `println()` statement and save the NonNumeric program. We will continue working with this program in the next section.

The **str()** Function

Processing's **str()** function can be used to convert a primitive-type value to a String value. The basic form of a call to the **str()** function is

```
str(value)
```

The parameter is

value   The primitive-type value to be converted to a String value

The type of value returned is **String** .

Return to the NonNumeric program. Suppose we wanted to output the concatenation of two integers, 12 and 34, producing 1234 as the result in the console. Enter the following program:

```
println(12 + 34);
```

When we run this program, the output to the console is

```
46
```

This is because both of the operands are of type int, so the + operation is interpreted as *addition*, not concatenation.

However, if we use the `str()` function to convert the operands to String values

```
str(12) + str(34)
```

then the result is the String value produced by the concatenation of the two String operands:

```
"1234"
```

Thus, the following statement,

```
println( str(12) + str(34) );
```

will produce the following output to the console:

```
1234
```

Also, remember that, if even *one* of the operands in a + operation is of type String, then the operation is interpreted as string *concatenation*, where both operands are simply put together to form a single String value. Thus, if we write instead

```
str(12) + 34
```

or

```
12 + str(34)
```

then the result of the concatenation operation will be the same String value:

```
"1234"
```

Save and close the NonNumeric program.

Simulating an Analog Thermometer

Start a new program by selecting **File > New**. Save this new program as **Thermometer**.

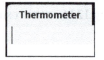

Let's write a program that uses some of the functions introduced in this chapter to simulate an analog thermometer. We'll start by declaring an int variable, assigning to it a random int value from 0 to 99, and displaying the randomly chosen value in the console:

```
int temperature;
temperature = int( random(100) );
println("Temperature: " + temperature);
```

When we run this program, we see console output like the following:

```
Temperature: 41
```

If we would like to output the degree symbol (°) as well, we can do so using the char() function. The decimal Unicode number for the degrees symbol is 176. Thus, we can add this symbol by writing

```
int temperature;
temperature = int( random(100) );
println("Temperature: " + temperature + char(176) );
```

Now, when we run this program, we see the degree symbol output to the console as well:

Temperature: 90°

If we would also like to display the letter F (for Fahrenheit), we can simply add

```
int temperature;
temperature = int( random(100) );
println("Temperature: " + temperature + char(176) + "F");
```

Now, when we run this program, we see console output like the following:

Temperature: 42°F

Next, let's draw a red rectangle with a length that is the same as the int value of our temperature variable. We'll need a canvas with a pixel width that is more than enough for our maximum temperature (99). Thus, let's insert the following call to the size() function:

```
int temperature;
temperature = int( random(100) );
println("Temperature: " + temperature + char(176) + "F");

size(150, 50);
```

Let's also set the fill color of the rectangle we will be drawing to red:

```
int temperature;
temperature = int( random(100) );
println("Temperature: " + temperature + char(176) + "F");

size(150, 50);
fill(255, 0, 0);
```

Now, let's draw the rectangle. We'll use 0 and 15 as the first two arguments of the rect() function so that the upper-left corner of the rectangle will be at column 0 and row 15 of the canvas:

```
rect(0, 15,
```

We'll use temperature as the third argument of the rect() function so that the length of the rectangle is the same number of pixels as the current temperature:

```
rect(0, 15, temperature,
```

And we'll use 20 as the fourth argument of the rect() function so that the rectangle is 20 pixels in height, completing our call to the `rect()` function:

```
rect(0, 15, temperature, 20);
```

Our program is now

```
int temperature;
temperature = int( random(100) );
println("Temperature: " + temperature + char(176) + "F" );

size(150, 50);
fill(255, 0, 0);
rect(0, 15, temperature, 20);
```

When we run our program, we see not only the temperature displayed in the console,

Temperature: 79°F

but also a red rectangle with a pixel length equal to the randomly chosen temperature:

Let's also display the temperature on the canvas at the end of this rectangle. Remember, Processing provides the **text()** function for drawing text on the canvas. The basic form of a call to the text() function is

text(*value, column, row*)

The value we wish to display on the canvas is the one stored in the temperature variable, so we'll set the first argument of the text() function to:

```
text(temperature,
```

The pixel column number at which to display the temperature is also the same as the value stored in the temperature variable, so we'll also set the second argument of the text() function to:

```
text(temperature, temperature,
```

And we can have the pixel row where the temperature is displayed be the same pixel row that we are using for the top of the rectangle if we set the third argument of the text() function to:

```
text(temperature, temperature, 15);
```

This completes the call to the `text()` function. Let's also use the `fill()` function to set the text color to black by inserting the following statement before our call to the `text()` function:

```
fill(0, 0, 0);
```

Thus, our program is now

```
int temperature;
temperature = int( random(100) );
println("Temperature: " + temperature + char(176) + "F" );
size(150, 50);

fill(255, 0, 0);
rect(0, 15, temperature, 20);
fill(0, 0, 0);
text(temperature, temperature, 15);
```

When we run this new version of our program, the random temperature that is generated and displayed in the console,

Temperature: 67°F

is also displayed in black text at the end of the rectangle:

Save the `Thermometer` program. We will use it again later in this book.

Special Functions for Input and Output with Dialog Boxes

Let's learn how to produce the kind of input and output dialog boxes that we are accustomed to seeing in software programs. For example, the following dialog box might be used to allow a user's name to be entered:

Start a new program by selecting **File > New**. Save this program as **Dialog**.

Processing does not itself contain functions for creating and using dialog boxes. However, because Processing is based on Java, we are able to make some use of Java's capability for dialog boxes. In order to do this, we just need to insert a special statement that is a reference to the Java "package" that contains these functions.

Unfortunately, this statement that we need to insert is a rather cryptic one. However, please don't be alarmed by it. You don't need to try to understand or memorize the syntax of this statement. It will only be used occasionally in this book, and on those occasions, it will always be written out for you. Just think of this statement as something like a necessary recipe ingredient that just happens to have a complicated name. Here is the single statement that we need to add to the first line of our program in order to be able to use dialog boxes:

```
import static javax.swing.JOptionPane.*;
```

Yes, it's a nuisance to have to type in such a cryptic statement. However, the resulting capability that we gain for creating and using dialog boxes is well worth the nuisance of adding this one line of code. Honestly, whenever we do use this statement, you can just type it in and then ignore it. Also, rest assured, there are *no other* Java-specific programming elements in this book. After all, this is a book about Processing, not about Java. Perhaps someday, a simpler way to create and use dialog boxes will be built into Processing.

Input Using a Dialog Box: The showInputDialog() Function
Processing does not provide a function for inputting a value from the console. However, now that we have added the above import statement, we are able to obtain input from the user through a dialog box by means of Java's showInputDialog() function.* The basic form of a call to the showInputDialog() function is

```
showInputDialog(prompt)
```

The parameter is

> *prompt*   The message displayed to the user. Type: Any, but usually a **String**.

The type of value returned by showInputDialog() is **String**.

For example, consider the following program:

```
import static javax.swing.JOptionPane.*;
showInputDialog("Enter your name:");
```

When we run this program, an input dialog box appears on the screen:

---

* In Java, a function is typically known as a "method."

We can then enter a name into the dialog box and press the OK button:

However, `showInputDialog()` is not a `void` function. Rather, it has a return type of `String`. This means that this function returns a `String` value to us—specifically, whatever `String` value is entered into the dialog box by the user.

Remember, whenever we call a function that returns a value, we need to *make immediate use* of that returned value. We have adopted the habit of doing this is by assigning the returned value to a variable of the same type:

```
variable = function(arguments);
```

Thus, in this case, let's define a variable of type `String` named `input` and assign the `String` value returned by the `showInputDialog()` function to this variable:

```
import static javax.swing.JOptionPane.*;

String input;
input = showInputDialog("Enter your name:");
```

The `String` value entered by the user is now stored in a `String` variable, `input`. As a result, we can make further use of this entered `String` value with this variable. For example, if we add the following statement to our program,

```
println("Hello, " + input + "!");
```

then the value entered by the user will be output to the console.

Our program is now

```
import static javax.swing.JOptionPane.*;
String input;
input = showInputDialog("Enter your name:");

println("Hello, " + input + "!");
```

When we run our program, the `String` value input by the user is read in from a dialog box,

and the output to the console is

```
Hello, Rebecca!
```

Save the `Dialog` program. We will continue using it in the next section.

Inputting an int or float Value

It is important to remember that the value entered by the user is always returned by the `showInputDialog()` function as a `String` value. Even if the user's input into the dialog *resembles* an `int` or `float` value, this entered value is nevertheless returned by the `showInputDialog()` function as a `String` value.

For example, let's change our `Dialog` program to the following program that prompts the user to enter an integer:

```
import static javax.swing.JOptionPane.*;

String input;
input = showInputDialog("Enter an integer to be squared:");
```

When we run this program, we see that the user can enter a value that *looks like* an integer:

However, as we have learned, the entered value returned by the `showInputDialog()` function is always a `String` value. In this particular case, the value returned is the `String` value `"8"` and not the `int` value 8 that it resembles. This is why we declared our `input` variable to be of type `String`. This creates a problem for us: as much as `"8"` looks like the `int` value 8, we can't do any mathematical operations with this returned `String` value that is stored in our `input` variable. Thus, we need to convert the `String` value `"8"` stored in our `input` variable into the corresponding `int` value 8 before we can square it.

Fortunately, as we learned earlier in this chapter, Processing's int() function will convert any String value that resembles an integer into its corresponding int value. This int result can be stored in an int variable that can then be used for mathematical operations. So, let's add the following two statements to our program:

```
import static javax.swing.JOptionPane.*;

String input;
input = showInputDialog("Enter an integer:");

int integer;
integer = int(input);
```

We can now do arithmetic operations with the int value stored in the integer variable. For example, we can calculate and display the square of the entered value by adding the following three statements to our program:

```
import static javax.swing.JOptionPane.*;

String input;
input = showInputDialog("Enter an integer to be squared:");

int integer;
integer = int(input);

int square;
square = integer * integer;

println(integer + " squared is: " + square );
```

When we run this program, the user can enter a value into an input dialog box,

and the square that corresponds to the user's entry is output to the console:

```
8 squared is: 64
```

Select **File > Save** to save the Dialog program. Then select **File > Save As** and re-save this program as Percent.

```
Percent    ▼

import s
```

In a similar manner, the **float()** function can be used to convert an appropriate String value entered by the user to the corresponding value of type float. For example, let's modify our current Percent program to the following:

```
import static javax.swing.JOptionPane.*;

String input;
input = showInputDialog("Enter a percentage:");

float percent;
percent = float(input);
float fraction;
fraction = percent / 100;

println(percent + "% is equivalent to: " + fraction);
```

Now, the user can enter a percentage into an input dialog box:

```
Input                                    ×

  ?   Enter a percentage:
      12.5

          OK       Cancel
```

In this case, "12.5" is the entered String value. Using the float() function, this String value is converted into a float value that is stored in the percent variable. This variable is then used to calculate the equivalent fraction that is displayed in the console:

```
12.5% is equivalent to: 0.125
```

Save and close the Percent program.

Output Using a Dialog Box: The showMessageDialog() Function

Select **File > New** to start a new program in Processing. By adding the import statement to this program,

```
import static javax.swing.JOptionPane.*;
```

we also gain access to the showMessageDialog() function, which displays a message to the user in a dialog box.

The basic form of a call to the showMessageDialog() function is

```
showMessageDialog(frame, message)
```

The parameter is

> **frame** This is a special keyword that we do not need to change.
>
> *message* The message displayed to the user. Type: Any, but usually a **String** .

The showMessageDialog() function is a **void** function. Thus, it does *not* return a value.
    To illustrate the use of this function, let's add the following statement to our current program.

```
import static javax.swing.JOptionPane.*;
showMessageDialog(frame, "Hello!");
```

When we run this program, we see the following message dialog box displayed to the user until the OK button is clicked:

Close this program without saving it. Reopen Then, reopen the Dialog program. We can display the squared integer in a dialog box instead of as output to the console simply by replacing the call to the println() function with a call to the showMessageDialog() function:

```
import static javax.swing.JOptionPane.*;

String input;
input = showInputDialog("Enter an integer to be squared:");

int integer;
integer = int(input);

int square;
square = integer * integer;

showMessageDialog(frame, integer + " squared is: " + square);
```

As before, when we run this program, the user enters an integer into the input dialog box:

But the square of this integer is now displayed to the user in a message dialog box:

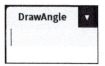

Save and close the `Dialog` program.

Interactive Example: Drawing an Angle

In this section, we will write a program that will draw an angle. Select **File > New** to start a new program. Save this program as **DrawAngle**.

```
DrawAngle  ▼
|
```

Let's start by drawing an unfilled black circle with a black stroke weight of 3 pixels and a diameter of 150 pixels. We'll position the center of this circle at the center of the canvas:

```
noFill();
stroke(0, 0, 0);
strokeWeight(3);
ellipse(width / 2, height / 2, 80, 80);
```

When we run this program, we now see the following:

Next, let's insert statements to declare an `int` variable named `degrees` to hold an angle measure. For now, we'll assign the value `45` to this variable. We'll also display the value of this variable in the console:

```
int degrees;
degrees = 45;

println("angle: " + degrees + char(176));

noFill();
stroke(0, 0, 0);
strokeWeight(3);
ellipse(width / 2, height /2, 80, 80);
```

When we run this program, we see the same circle drawn as before, but now we also see the angle measure output to the console:

angle: 45°

Instead of drawing a circle, let's now draw an arc using our chosen angle. In order to use the `arc()` function, we need to convert our angle measure to radians using the `radians()` function. Also, we'll use the `PIE` option for the `arc()` function so that a fully outlined pie wedge shape is drawn. Our program is now:

```
int degrees;
degrees = 45;
println("angle: " + degrees + char(176));

noFill();
stroke(0, 0, 0);
strokeWeight(3);
arc(width / 2, height /2, 80, 80, 0, radians(degrees), PIE);
```

When we run our program, we see the same console output as before:

angle: 45°

However, we now see the following representation of this angle drawn upon the canvas:

To draw the angle in the *counterclockwise* way that is more commonly used in geometry, we can start instead at the *negative* of the angle measure and draw back to angle measure 0. After we make this change, our program is now:

```
int degrees;
degrees = 45;
println("angle: " + degrees + char(176));

noFill();
stroke(0, 0, 0);
strokeWeight(3);
arc(width / 2, height /2, 80, 80, -radians(degrees), 0, PIE);
```

Now, when we run our program, we see the same console output as before:

angle: 45°

However, we now see the following representation of the angle drawn on the canvas:

Let's also draw a white arc using the same arguments as before but without the PIE option. We can do so by adding two statements to our program:

```
int degrees;
degrees = 45;
println("angle: " + degrees + char(176));

noFill();
stroke(0, 0, 0);
strokeWeight(3);
arc(width / 2, height /2, 80, 80, -radians(degrees), 0, PIE);

stroke(255, 255, 255);
arc(width / 2, height / 2, 80, 80, -radians(degrees), 0);
```

When we run this program, we see the same console output as before:

angle: 45°

However, we now see the following representation of the angle drawn on the canvas:

Next, let's modify the program so that the user can enter the angle measure. For this, we'll need to insert this import statement at the beginning of our program so that we can generate an input dialog:

```
import static javax.swing.JOptionPane.*;
```

Next, let's input our angle. As we have learned, the showInputDialog() function reads in and returns the user's input as a String value, but we can convert that String value to an int value using the int() function and assign the result to our degrees variable. After we do this, our program is now

```
import static javax.swing.JOptionPane.*;

String input;
input = showInputDialog("Enter angle in degrees:");

int degrees;
degrees = int(input);
println("angle: " + degrees + char(176));

noFill();
stroke(0, 0, 0);
strokeWeight(3);
arc(width / 2, height /2, 80, 80, -radians(degrees), 0, PIE);

stroke(255, 255, 255);
arc(width / 2, height / 2, 80, 80, -radians(degrees), 0);
```

When we run our program, the user is now prompted for an angle:

The entered angle measure is displayed back to the user in the console,

angle: 135°

and a representation of the angle is drawn on the canvas:

Save the `DrawAngle` program. We will return to it in the next chapter.

Summary

In this chapter, we learned much more about how functions operate in Processing.

- We learned about `void` functions, which do not return a value.

- We learned about a number of functions that do return a value, such as the `sqrt()` and `pow()` functions.

- We acquired the ability to generate random values using the `random()` function.

- We covered functions such as the `round()` function, which converts from the `float` to the `int` type.

- We converted values from another type into an `int` or `float` type of value using the `int()` and `float()` functions, respectively.

- We learned how to convert `int` values to `char` and `String` types using the `char()` and `str()` functions, respectively.

- We learned about inputting information from the user by means of a dialog box using Java's `showInputDialog()` function.

- We learned how to generate output to a dialog box using Java's `showMessageDialog()` function.

Exercises

1) Write a program that will display the following output to the console:

This

is

Processing.

2) Write a program to do the following:

   a) Declare an `int` variable named `total`.

   b) Initialize the `total` variable to `10`.

   c) Use the `total` variable in a call to the `println()` function to display the variable's value to the console in the following format:

      total: 10

3) Write a program to do the following:

   a) Declare an `int` variable named `amount`.

   b) Initialize the `amount` variable to `50`.

   c) Use the `amount` variable in a call to the `println()` function to display the variable's value to the console in the following format:

      50 is the amount.

4) Write a program to do the following:

    a) Declare an `int` variable named `students`.

    b) Initialize the `students` variable to `75`.

    c) Use the `students` variable in a call to the `println()` function to display the variable's value to the console in the following format:

       There are 75 students in all.

5) Write a program that draws your first name on the canvas with the `text()` function, using whatever font color, font size, and canvas size you choose.

6) Write a program to do the following:

    a) Declare a variable named `squareRoot` that can hold a number with decimal places.

    b) Declare a variable named `square` that can hold an integer.

    c) Assign to `square` the value `1681`.

    d) Calculate the square root of `square` and store the result in `squareRoot`.

    e) Display the value of `square` and the value of `squareRoot` in the console.

    f) Your console output should be

       square: 1681

       square root: 41.0

7) Write a program to do the following:

    a) Declare a variable named `base`.

    b) Initialize `base` to `2`.

    c) Declare a variable named `exponent`.

    d) Initialize `exponent` to `5`.

    e) Declare a variable named `solution`.

    f) Use the `pow()` function to raise `base` to the power of `exponent`.

    g) Use all three variables—`base`, `exponent`, and `solution`—in a call to the `println()` function that displays the result in the console in the following format:

       2 to the power 5 is 32.0

8) Write a program to do the following:

a) Declare a variable named `side` that can hold an integer.

b) Set the value of `side` to `15`.

c) Declare a variable named `volume` that can hold a number with decimal places.

d) Use `side` and the `pow()` function to calculate the volume of cube with a side length equal to the value of `side` and store the result in `volume`.

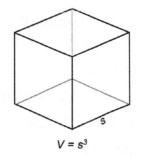

$$V = s^3$$

e) Display the value of side and the value of volume in the console. Your output should be

```
cube side length: 15
cube volume: 3375.0
```

9) Write a program to do the following:

a) Declare a variable named `radius` to store the radius of a circle.

b) Initialize `radius` to `10`.

c) Declare a variable named `area`.

d) Use the `radius` variable, Processing's built-in constant `PI`, and the `pow()` function to assign the area of the circle to `area`.

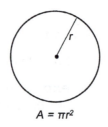

$$A = \pi r^2$$

e) Use `radius` and `circumference` in a call to the `println()` function to display the following in the console:

```
Area of a circle with radius 10 is 314.15927
```

10) Write a program to do the following:

a) Declare a variable named `radius` that can store a number with decimal places.

b) Set the value of `radius` to `10`.

c) Declare a variable named `volume` that can hold a number with decimal places.

d) Use Processing's `PI` constant and the `pow()` function to calculate the volume of a sphere that has a radius equal to the value of `radius` and assign the result to `volume`.

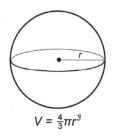

$$V = \tfrac{4}{3}\pi r^3$$

e) Display in the console "`radius:` ", followed by the value of `radius`.

f) Display in the console "`sphere volume:` ", followed by the value of `volume`.

Your console output should be

```
radius: 10.0
sphere volume: 4188.7905
```

Hint: Watch out for integer division.

11) Write a program to do the following:

a) Declare a variable named `radius` that can hold an integer.

b) Set the value of `radius` to `25`.

c) Declare a variable named `altitude` that can hold an integer.

d) Set the value of `altitude` to `100`.

e) Assume that these two variables describe the dimensions of a cylinder.

f) Declare a variable named `volume` that can hold a number with decimal places.

g) Use `radius`, `height`, Processing's `PI` constant, and the `pow()` function to calculate the volume of the cylinder and assign the result to `volume`.

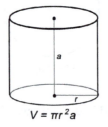

$$V = \pi r^2 a$$

h) Display the values of `radius`, `height`, and `volume` to the console.

Your console output should be

```
cylinder radius: 25
cylinder altitude: 100
cylinder volume: 196349.55
```

12) For each of the following values, write code that uses the `round()` function to convert the given value to an integer and displays the integer in the console.

Example

```
println( round(3.25) ) ;
```

5.1
5.7
5.5
5.49
5.4999999

13) For each of the following values, write code that uses the int () function to convert the given value to an integer and displays the integer in the console.

Example

```
println( int(3.25) ) ;
```

```
1.7
1.9999
5.8
6.1
```

14) Write a program to do the following:

a) Declare a variable named age and initialize it to 17.5.

b) Declare a variable named integerAge.

c) Use the int() function to convert age to an integer and assign the result to integerAge.

d) Use age and integerAge in a call to the println() function that displays the following in the console:

```
The integer portion of 17.5 is 17
```

15) Write a program to do the following:

a) Declare an int variable named total and initialize it to 105.

b) Declare an int variable named count and initialize it to 10.

c) Declare a float variable named average.

d) To avoid integer division, use the float() function when dividing total by count to calculate the average and assign the result to average.

e) Use average in a call to the println() function that displays the following to the console:

```
The average is: 10.5
```

16) Write a program that picks a random float value between 0 and 50 and displays the result in the console.

17) Write a program to do the following:

   a) Declare a variable named `randomNumber`.

   b) Use the `random()` function to assign a random value between 1 and 100 to `randomNumber`.

   c) Declare an `int` variable named `randomInteger`.

   d) Use the `round()` function to convert `randomNumber` to an integer and assign the result to `randomInteger`.

   e) Use `randomNumber` and `randomInteger` in a call to the `println()` function that displays the values of these two variables in the console in the following format:

   ```
   27.225977 rounded is 27
   ```

18) Write a program that, each time it is run, draws a 10-pixel by 10-pixel square at a random pixel column along the top of a 100-pixel by 100-pixel canvas.

   Example

19) Write a program that, each time it is run, draws a line segment from the upper-left corner of a 150-pixel by 200-pixel canvas to a random pixel location on that canvas.

   Example

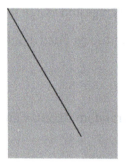

20) Write a program to do the following:

a) Set the size of the canvas to 150 pixels by 200 pixels.

b) Declare a variable named `column` that can store a number with decimal places.

c) Assign to `column` a random value from 0 up to (but not including) the width of the canvas. (Use Processing's built-in `width` variable for this.)

d) Declare a variable named `row` that can store a number with decimal places.

e) Assign to `row` a random value from 0 up to (but not including) the height of the canvas. (Use Processing's built-in `height` variable for this.)

f) Draw a circle with a diameter of 20 pixels on the canvas at the chosen column and row location.

g) Draw a line segment from the upper-left corner of the canvas to the chosen column and row location.

h) Your canvas should look something like

i) Use the `println()` and `int()` functions to display an integer version of the value of `column` in the console.

j) Use the `println()` and `int()` functions to display an integer version of the value of `row` in the console.

k) Your console output should be something like

```
column: 99
row: 151
```

21) Write a program to do the following:

a) Use `showMessageDialog()` to display a dialog that says "Greetings!"

b) Your output should be

22) Write a program that reads in a user's favorite fruit,

> Input   ✕
>
> **?**   **Enter your favorite fruit:**
> cherries
>
> OK   Cancel

and outputs it to the console.

> **You love cherries!**

23) Modify the preceding program so that the user's favorite fruit is output in a message dialog box.

> Message   ✕
>
> ⓘ   You love cherries!
>
> OK

24) Write a program that reads in a person's current age, converts it to an integer, and displays in the console the age that the person will be *next* year.

> Input   ✕
>
> **?**   **Enter your age:**
> 19
>
> OK   Cancel

> **Your age next year: 20**

25) Modify the preceding program so that the user's age is output in a message dialog box.

> Input   ✕
>
> **?**   **Enter your age:**
> 19
>
> OK   Cancel

> Message   ✕
>
> ⓘ   Your age next year: 20
>
> OK

CHAPTER 4

Conditional Programming with `if`

Up to now, all our programs have been what are sometimes known as **straight-line** programs, because all of the statements in such a program are simply executed in the order in which they are listed, from first to last.

In this chapter, we wish to look at statements that are **conditional** in that they involve one or more actions that are performed only when a certain *condition* is true. Such conditional statements are also known as examples of **selection**, because they specify actions that are performed only *selectively*.

Recipe Analogy

A cooking recipe can provide a helpful analogy when we think about what a computer program is. Like a recipe, a program involves performing a certain set of individual actions in a particular sequence, from beginning to end, to achieve a specific end result. In computer programming, such a sequence of actions that produces a desired end result is also known as an **algorithm**.

However, there are times when we might wish to perform one or more of the individual actions in such a sequence *selectively*—that is, only *if* a certain *condition* is true. Consider our recipe analogy. Butter can be purchased salted or unsalted. Thus, a recipe might include a statement such as

If you are using <u>un</u>salted butter, add ¼ teaspoon of salt.

According to this instruction, we do *not always* want to add this ¼ teaspoon of salt to the recipe. Rather, this action is *conditional*. Specifically, *if* the condition is *true* (the butter is indeed unsalted), then we *do* want to perform the action of adding salt. However, if our condition is *false* (the butter is *not* unsalted, but rather is salted butter), then we *do not* perform the action of adding salt.

Such a conditional action is sometimes represented as a **flowchart**:

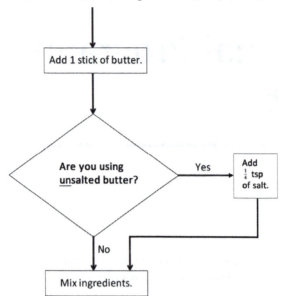

Similarly, a computer program may include *conditional* statements. Such statements are performed *selectively*, only if a certain *condition* is true.

The Basic if Statement

Most programming languages provide a very powerful feature for conditional action: the **if statement**. In Processing, the basic **if** statement is typically written in the form

```
if (condition)
{
    Statement(s) to perform if the condition is true
}
```

Unlike the other statements we have written so far, the if statement is usually written as more than one line of code, and yet it is still considered to be a single statement.

Let's take a closer look at the items of information that we need to supply for the general form of the if statement. First, an if statement needs a test **condition**.

```
if (condition)
```

This test condition is usually a *comparison* that will always turn out to be either **true** or **false**. This kind of true-or-false comparison is also known as a **Boolean** expression.* In Processing, the basic operators for comparison, known as the **relational operators**, are the following:

>   Greater than

<   Less than

---

* Boolean expressions are named after George Boole (1815–1864), an English mathematician who, in addition to many other accomplishments, developed methods for applying algebra to logic.

>=   Greater than or equal to

<=   Less than or equal to

!=   Not equal to

==   Equal to

Second, we need to specify one or more statements that will be performed *if*—and only if—the condition is *true*. We will enclose any such statement(s) in a pair of curly braces, creating what is known as a **block** of statements.

```
if (condition)
{
    Statement(s) to perform if the condition is true
}
```

Basic if Statement Example: Rolling a Die Game

Let's return to the die-rolling Roll program from Chapter 3. Currently, it consists of the following statements:

```
int roll;
roll = int( random(1, 7) );
println("Roll: " + roll);
```

When we run this program, we see the following kind of result in the console:

Roll: 2

Now, suppose we are instead playing a game in which the player wins only *if* the number rolled is *greater than* 3. For this, our test condition can be

roll > 3

This test condition is comprised of a *comparison*. Each time we run this program, this comparison will always turn out to be either *true* or *false*. These are only two possibilities. One possibility is that the roll is *greater than* 3 (i.e., the roll is a 4, 5, or 6). In this case, the comparison in our condition is *true*. The other possibility is that the roll is *not greater than* 3. Instead, the roll is less than or equal to 3 (i.e., the roll is a 1, 2, or 3). In this case, the comparison in our condition is *false*.

roll > 3   ⚃⚄⚅ **true**

⚀⚁⚂ **false**

According to the general form of the `if` statement, we enclose our condition in a pair of parentheses. Thus, for the current program, we will start our `if` statement by adding the following line of code to our program:

```
int roll;
roll = int( random(1, 7) );
println("Roll: " + roll);

if (roll > 3)
```

Next, we need to specify the one or more statements that will be executed *if*—and only if—the condition is *true*.* We will enclose these statements in a pair of curly braces and indent them, creating a program *block*:

```
if (roll > 3)
{
    Statement(s) to perform if roll > 3
        ⋮
}
```

We now need to specify what it is that we want to do *if* the roll is indeed *greater than* 3. Let's display a message to the console announcing that the player has won the game. For this, we can insert the following statement:

```
int roll;
roll = int( random(1, 7) );
println("Roll: " + roll);

if (roll > 3)
{
    println("You won!");
}
```

Now, when we run this program, if the roll is *greater than* 3, then the condition is *true*, and the message about winning the game is indeed displayed:

```
Roll: 4
You won!
```

However, if the roll is *not greater than* 3 (i.e., is less than or equal to 3), then the condition is *false*, and the reminder about winning the game is not displayed:

---

* Technically, if there is only *one* statement to be executed, then curly braces are not required. Thus, we could write:
```
if (roll > 3)
    println("You won!");
```
However, both for consistency and to allow for the later addition of other statements, we will use curly braces in *every* `if` statement in this book.

```
Roll: 3
```

We can visualize the logic of this if statement with the following flowchart:

Save the Roll program. We will continue working with this program in the next section of this chapter.

Caution: No Semicolon at the End of the First Line of an if Statement

Always make sure that there is *no* semicolon at the end of the first line of an if statement.

if (roll > 3)[;]

↑
Do **not** add a
semicolon here!

We are so accustomed to adding a semicolon to the end of single-line statements that adding a semicolon to the first line of an if statement is a very easy mistake to make. However, this can be a very difficult error to find, because it does not cause an error message. For example, if we accidentally write the following, then no error message results:

```
if (roll > 3);
{
  println("You won!");
}
```

However, this is a logical error because we have now created an if statement that actually consists only of the following:

```
if (roll > 3);
```

This is an if statement that takes *no action*, even if the condition is true! In other words, we have essentially given the following instruction to Processing: "If the roll is greater than 3, then don't do anything." Also, because of this misplaced semicolon, the statement block following the if statement no longer has a condition attached to it. Instead, Processing essentially views this statement block as:

```
{
  println("You won!");
}
```

As a result, this statement block will now be performed *every time* the program is run, even when the roll is *not greater than* 3. For example,

```
Roll: 1
You won!
```

Thus, we need to be careful to avoid adding a semicolon at the end of the first line of an if statement. (Even experienced programmers occasionally make this mistake.) When you do, you will probably find that it is easy to keep *overlooking* this mistake as you try to figure out why your program is not producing the correct results. So, here's a good habit to adopt: whenever an if statement is not operating correctly, always check to see if you have accidentally added a semicolon to the end of the first line.

Check again to make sure that there is no semicolon at the end of the first line of your if statement in our current program. Then, save the Roll program. We will continue working with this program in the next section.

Checking Equality: int Values

Now, suppose that we wish to change our game so that we win only if we roll a 6. In order to achieve this result, we need to change our if statement's condition to

```
if (roll == 6)
```

After making this change, our program is now

```
int roll;
roll = int( random(1, 7) );
println("Roll: " + roll);

if (roll == 6)
{
  println("You won!");
}
```

As a result, the condition in our if statement will now be true only if the value of roll is 6. Otherwise, the condition will be false.

$$roll == 6 \quad \boxed{\vdots} \textbf{ true}$$

$$\boxed{\cdot} \text{ or } \boxed{\cdot\cdot} \text{ or } \boxed{\cdot\cdot\cdot} \text{ or } \boxed{::} \text{ or } \boxed{:\cdot:} \textbf{ false}$$

Thus, when we run our program, we now win only if the roll is a 6:

```
Roll: 6
You won!
```

Notice that we use *two* equals signs (==) as the operator for comparing two items to see if they are equal. In Processing, the *double* equals sign operator is used to compare for equality because, as we have learned, the *single* equals sign is the *assignment operator*. It is easy to make the mistake of typing a *single* equals sign when a double equals sign is needed. For example,

```
if (roll = 6)
{
  println("You won!");
}
```

Fortunately, if we accidentally do this, Processing underlines this condition,

```
if (roll = 6)
```

and, in this case, gives us the following error message in the Message Area:

Type mismatch, "int" does not match with "boolean"

This error message isn't entirely clear. However, we have seen that a type mismatch often results in a failed assignment statement, and this is precisely the problem we have here: we have used the assignment operator (=) instead of the equality comparison operator (==). Remember:

- A *single* equals sign (=) is used for *assigning* a value to a variable.

- A *double* equals sign (==) is used for *comparing* two values.

Save the Roll program. We will return to this program later in this chapter.

Basic if Statement: A Graphical Example

Let's create a graphical program that uses the basic if statement. Start a new program by selecting **File > New**. Save this program as **RandomDiameter**.

Let's set the size of the canvas to 200 pixels by 200 pixels.

```
size(200, 200);
```

Next, let's declare an `int` variable named `diameter` and assign it a random value between 5 pixels and the width of the canvas. Let's also display this value in the console:

```
size(200, 200);

int diameter;
diameter = int( random(5, width) );
println("Diameter: " + diameter);
```

We'll now add the following statement to draw a circle that has the chosen diameter and its center at the horizontal and vertical center of the canvas:

```
size(200, 200);

int diameter;
diameter = int( random(5, width) );
println("Diameter: " + diameter);

ellipse(width / 2, height / 2, diameter, diameter);
```

When we run this program, we do indeed see a circle with a random diameter drawn with its center positioned at the horizontal and vertical center of a canvas that is 200 pixels by 200 pixels:

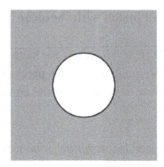

We also see the randomly chosen diameter displayed in the console:

```
Diameter: 85
```

Now, let's set the fill color of the circle to red whenever the diameter chosen for the circle is greater than half of the width of the canvas (i.e., greater than 100 pixels). For this, we can insert the following `if` statement:

```
if (diameter > (width / 2) )
{
  fill(255, 0, 0); // fill red
}
```

After making this change, our program is now

```
size(200, 200);

int diameter;
diameter = int( random(5, width) );
println("Diameter: " + diameter);

if (diameter > (width / 2) )
{
  fill(255, 0, 0); // fill red
}

ellipse(width / 2, height / 2, diameter, diameter);
```

Now, when we run our program, if the chosen diameter is, for example,

Diameter: 137

then the chosen diameter is *greater* than half of the width of the canvas (i.e., greater than 100 pixels), and the condition in our `if` statement is `true`. As a result, our call to the `fill()` function is performed, setting the fill color of the circle drawn to red:

On the other hand, if the chosen diameter is , for example,

Diameter: 15

then the chosen diameter is *not greater* than half of the width of the canvas (i.e., less than or equal to 100 pixels), and the condition in our `if` statement is `false`. As a result, our call to the `fill()` function is *not* performed, so the circle drawn has the default fill color, white:

We can visualize the logic of this `if` statement with the following flowchart:

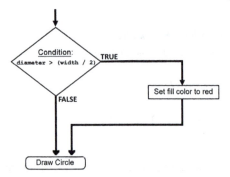

Save this `RandomDiameter` program. We will return to it later in this chapter.

The `if-else` Statement

We have seen that the `if` statement allows us to specify a condition and perform certain individual actions only when that condition is *true*. An `if-else` statement allows us to specify also the actions that we would like to perform in the *alternative* scenario: when our condition is not true but is instead *false*.

The basic structure of the **if-else** statement is typically

```
if (condition)
{
    Statement(s) to perform if the condition is true
}
else
{
    Statement(s) to perform if the condition is false
}
```

This is a single Processing statement that now has *two* **clauses**: an `if` clause and an `else` clause.

Let's return to the `if` statement in our example of simulating the rolling of a die that we created our `Roll` program. This program currently consists of the following:

```
int roll;
roll = int( random(1, 7) );
println("Roll: " + roll);

if (roll == 6)
{
  println("You won!");
}
```

When the value of `roll` is equal to 6, the condition of this `if` statement is true. Otherwise, the value of `roll` is equal to 1, 2, 3, 4, or 5, and the condition is false.

$$\texttt{roll == 6}$$

⊞ **true**

⊡ or ⊡ or ⊡ or ⊡ or ⊡ **false**

Currently, when we run our program, a message announcing that the player has won is displayed only when the roll is a 6:

```
Roll: 6
You won!
```

Otherwise, the roll is a 1, 2, 3, 4, or 5, and the player has lost. However, in this case, *no* message is displayed following the roll:

```
Roll: 2
```

We should also display a message following the roll when the player loses. For this, we will add an `else` clause:

```
int roll;
roll = int( random(1, 7) );
println("Roll: " + roll);

if (roll == 6)
{
  println("You won!");
}
else
{

}
```

An `else` clause specifies the alternative actions that will be performed whenever the condition is *false*.* In our current program, we would like to display a message indicating that the player has lost. Thus, our `else` clause can be

```
int roll;
roll = int( random(1, 7) );
println("Roll: " + roll);

if (roll == 6)
{
  println("You won!");
}
```

___
* As in the case of the basic `if` statement, the curly braces are technically unnecessary when a *single* statement follows the `else` keyword. However, for the sake of consistency and to make it easier to add additional statements, we will always use curly braces when writing an `else` clause.

```
else
{
  println("Sorry, you lost.");
}
```

Now, when we run our program, the program displays the same message as before when a 6 is rolled:

```
Roll: 6
You won!
```

However, because of our added `else` clause, a message is also displayed indicating that the player has lost when something other than a 6 is rolled. For example,

```
Roll: 1
Sorry, you lost.
```

To summarize, the `if` clause specifies the statements to be performed whenever the condition is *true*

```
if (roll == 6)
{
  println("You won!");
}
else
{
  println("Sorry, you lost.");
}
```

and the `else` clause specifies the statements to be performed whenever the condition is *false*:

```
if (roll == 6)
{
  println("You won!");
}
else
{
  println("Sorry, you lost.");
}
```

We can visualize the logic of this `if-else` statement using the following flowchart:

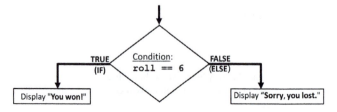

You might find it useful to paraphrase the word `else` as "otherwise" when working with `if-else` statements: "*If* the condition is true, then perform these actions; *otherwise*, perform these other actions."

Save the `Roll` program. We will return to it later in this chapter.

Using the `if-else` Statement: Classifying Ages

Select **File > New** to start a new program. Save this program as **Ages**.

Enter the following statements into this program:

```
int age;
age = int( random(1, 30) );
println("Age: " + age);
```

When we run this program, we see that a randomly chosen integer age from 1 to 29 is displayed in the console:

Age: 23

Now, suppose that we would like to classify the chosen age in terms of whether it is the age of an *adult* or is the age of a minor. Essentially, we are asking this question:

"Is the age 18 or older?"

If the answer is yes, then the age is classified as that of an adult. Otherwise, the answer is no, and the age is classified as that of a minor. We can create the logic of asking this question in our program by using the following `if-else` statement:

```
int age;
age = int( random(1, 30) );
println("Age: " + age);

if (age >= 18)
{
  println("Adult");
}
else
{
  println("Minor");
}
```

We can visualize the logic of this `if-else` statement using the following flowchart:

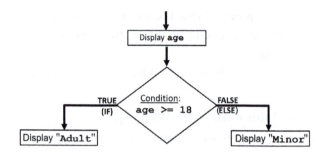

When we run this program, we see from the console output that the age chosen is now correctly classified as either the age of an adult,

```
Age: 27
Adult
```

or the age of a minor,

```
Age: 2
Minor
```

Save the Ages program. We will return to it later in this chapter.

Using the if-else Statement: A Graphical Example

Let's return to the RandomDiameter program. Currently, the if statement in this program sets the fill color to red only if the randomly selected diameter is *greater than* half of the width of the canvas:

```
size(200, 200);

int diameter;
diameter = int( random(5, width));
println("Diameter: " + diameter);

if (diameter > (width / 2) )
{
  fill(255, 0, 0);
}

ellipse(width / 2, height / 2, diameter, diameter);
```

Otherwise, if the chosen diameter is *not greater than* half of the width of the canvas (i.e., not greater than 100 pixels),

```
Diameter: 67
```

then the fill color of the circle drawn is white, Processing's default fill color:

Let's add an `else` clause so that the fill color is set to blue instead of the default fill color white whenever the randomly selected diameter is not greater than half of the width of the canvas. Once again, we can visualize this conditional logic as a flowchart:

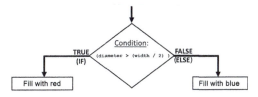

We'll change our program to

```
size(200, 200);

int diameter;
diameter = int( random(5, width));
println("Diameter: " + diameter);

if (diameter > (width / 2) )
{
  fill(255, 0, 0);
}
else
{
  fill(0, 0, 255);
}

ellipse(width / 2, height / 2, diameter, diameter);
```

When we run this modified program, if the selected diameter is *greater than* half the width of the canvas (i.e., greater than 100 pixels),

Diameter: 177

then, as before, the fill color of the circle drawn is red:

However, if the selected diameter is *not greater than* half the width of the canvas (i.e., *not greater than* 100 pixels),

```
Diameter: 66
```

then the `else` clause is performed instead, setting the fill color of the circle drawn to blue:

In our current program, let's also use end-of-line comments to clarify the fill color that is chosen.

```
if (diameter > (width / 2) )
{
  fill(255, 0, 0);        // red
}
else
{
  fill(0, 0, 255);        // blue
}
```

Remember, we may add comments like this because anything from two slashes (//) to the end of a line will be *ignored* by Processing when we run our program.

`fill(255, 0, 0);` `// red` ← **ignored by Processing**

Save and close the `RandomDiameter` program.

The `if-else-if-else` Statement

There are times when we might wish to make use of another option offered for the `if` statement: `else if` clauses. An `if-else-if-else` statement allows us to test for multiple conditions that are *mutually exclusive*. In other words, only *one* of the conditions will ever be true at any given time. The simplest form of the `if-else-if-else` statement is

```
if (condition-1)
{
  Statement(s) to perform if condition-1 is true
}
```

```
else if (condition-2)
{
    Statement(s) to perform if condition-2 is true
}
else
{
    Statement(s) to perform if neither
    of the above conditions is true
}
```

However, we are actually allowed to add as many `else if` clauses as we would like:

```
if (condition-1)
{
    Statement(s) to perform if condition-1 is true
}
else if (condition-2)
{
    Statement(s) to perform if condition-2 is true
}
    ⋮
else if (condition-n)
{
    Statement(s) to perform if condition-n is true
}
else
{
    Statement(s) to perform if none of the above conditions is true
}
```

Thus, the `if-else-if-else` statement offers us the opportunity to test for what could even be an extensive list of mutually exclusive possibilities. This form of selection will become clearer after we look at some examples in the next several sections.

Rock–Paper–Scissors Example: if-else-if-else

Start a new program by selecting **File** > **New**. Save this new program as **Rock**.

Let's create a simple version of the "rock–paper–scissors" game.

We'll start by picking a random integer from 1 to 3, storing it in an `int` variable, and displaying it in the console:

```
int integer;
integer = int( random(1, 4) );
println("integer is: " + integer);
```

When we run this program, we see that a 1, 2, or 3 is picked at random:

integer is: 1

or

integer is: 2

or

integer is: 3

If the randomly chosen integer is 1, let's have the computer pick "rock." If the randomly chosen integer is 2, let's have the computer pick "paper." If the randomly chosen integer is 3, let's have the computer pick "scissors." For to create this logic, we could add the following three if statements:

```
int integer;
integer = int( random(1, 4) );
println("integer is: " + integer);

if (integer == 1)
{
  println("Computer picks: Rock!");
}

if (integer == 2)
{
  println("Computer picks: Paper!");
}

if (integer == 3)
{
  println("Computer picks: Scissors!");
}
```

Now, when we run our program, if the randomly chosen integer is 1, then the computer picks "rock:"

integer is: 1
Computer picks Rock!

If the randomly chosen integer is 2, then the computer picks "paper:"

integer is: 2
Computer picks Paper!

If the randomly chosen integer is 3, then the computer picks "scissors:"

```
integer is: 3
Computer picks Scissors!
```

However, what is undesirable about this version of our program is that Processing will always perform *all three* if statements, even when it is not necessary to do so. For example, whenever the randomly chosen integer is 1, Processing performs the first if statement, and the computer picks "rock." But Processing will also go on to test the conditions in the second and third if statements, even though these conditions can never be true when the integer chosen is 1.

```
          true!
if ((integer == 1))
{
  println("Computer picks: Rock!");
}


          false
if ((integer == 2))       ← test is performed even though it is not necessary
{
  println("Computer picks: Paper!");
}


          false
if ((integer == 3))       ← test is performed even though it is not necessary
{
  println("Computer picks: Scissors!");
}
```

Likewise, whenever the randomly chosen integer is 2, Processing will still unnecessarily go on to test the condition in the third if statement, even though this condition can never be true when the integer chosen is 2.

```
          false
if ((integer == 1))
{
  println("Computer picks: Rock!");
}


          true!
if ((integer == 2))
{
  println("Computer picks: Paper!");
}
```

```
          false
if (integer == 3)        ← test is performed even though it is not necessary
{
  println("Computer picks: Scissors!");
}
```

Only *one* of these conditions can be true at any given time. Thus, as soon as one of these conditions is determined to be true, we would like Processing to *skip* any remaining conditions. Fortunately, this is precisely what happens if we use an if-else-if-else statement:

```
if (integer == 1)
{
  println("Computer picks: Rock!");
}

else if (integer == 2)
{
  println("Computer picks: Paper!");
}

else
{
  println("Computer picks: Scissors!");
}
```

Now, whenever we run this program and the randomly chosen integer is 1, Processing tests the condition in the if clause and determines it to be true, so the statement associated with the if clause is performed: computer picks "rock." However, any else if or else clauses following the if clause will *not* be performed.

```
          true!
if (integer == 1)
{
  println("Computer picks: Rock!");
}
else if (integer == 2)
{
  println("Computer picks: Paper!");
}
else
{
  println("Computer picks: Scissors!");
}
```

On the other hand, whenever the randomly chosen integer is 2, the condition in the if clause is tested first and determined to be false, so the statement it contains is not performed.

Processing then goes on to test the condition in the `else if` clause and determines it to be true, so the statement it contains is performed, and the computer chooses "paper." However, any remaining `else if` or `else` clauses are skipped over. In this particular case, it is the `else` clause that is not performed.

```
             false
if (integer == 1)
{
   println("Computer picks: Rock!");
}
                true!
else if (integer == 2)
{
   println("Computer picks: Paper!");
}

else
{
   println("Computer picks: Scissors!");
}
```

Such skipping over unnecessary conditions is sometimes known as **short-circuit evaluation**.

As in an `if-else` statement, the **else** clause of an `if-else-if-else` statement defines the *default* actions. These are the statements that are performed only when *all* the preceding conditions of the `if` clause and any `else if` clauses have already been determined to be *false*. In our program, if the randomly chosen integer has already been determined to be neither 1 nor 2, then the only remaining possibility is for the integer to be **3**, and in this case, the computer should pick "**scissors**." This is why the `else` clause was used for the "scissors" case.

```
             false
if (integer == 1)
{
   println("Computer picks: Rock!");
}
                false
else if (integer == 2)
{
   println("Computer picks: Paper!");
}

else
{
   println("Computer picks: Scissors!");
}
```

Let's add a comment to clarify the default logic of our else clause:

```
else // integer is 3
{
  println("Computer picks: Scissors!");
}
```

Our program is now

```
int integer;
integer = int( random(1, 4) );
println("integer is: " + integer);

if (integer == 1)
{
  println("Computer picks: Rock!");
}
else if (integer == 2)
{
  println("Computer picks: Paper!");
}

else // integer is 3
{
  println("Computer picks: Scissors!");
}
```

When we run this program, we do indeed get the output to the console that we expect:

```
integer is: 1
Computer picks: Rock!
```

or

```
integer is: 2
Computer picks: Paper!
```

or

```
integer is: 3
Computer picks: Scissors!
```

We can visualize the logic of this if-else-if-else statement with the following flowchart:

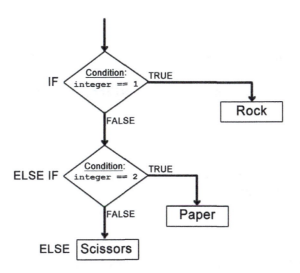

Save this Rock program. We will return to this example in the next section.

Using the `if-else-if-else` Statement: A Graphical Example

We can easily alter our rock–paper–scissors program so that it instead draws a red square, blue circle, or yellow triangle.* Return to the Rock program. Select **File > Save As** and resave this program as **Shapes**.

This Shapes program will randomly draw a yellow triangle, a red square, a blue circle, or a green triangle if we make the following changes to it:

```
int integer;
integer = int( random(1, 4) );
println("integer is: " + integer);

noStroke();

if (integer == 1)
{
  println("Red Square");
  fill(255, 0, 0);
  rect(0, 0, width, height);
}
```

---

* In 1923, Russian painter Wassily Kandinsky (1866–1944) did a survey at the famous Bauhaus art school in Germany and discovered a surprising tendency among his colleagues to assign these three colors to these three shapes. Believing this was evidence of a universal correspondence between these shapes and colors, Kandinsky used them in many of his works.

```
else if (integer == 2)
{
  println("Blue Circle");
  fill(0, 0, 255);
  ellipse(width / 2, height / 2, width, height);
}

else // integer is 3
{
  println("Yellow Triangle");
  fill(255, 255, 0);
  triangle(width / 2, 0, 0, height, width, height);
}
```

Now, when we run this program, the output to the console and canvas is

```
integer is: 1
Red Square
```

or

```
integer is: 2
Blue Circle
```

or

```
integer is: 3
Yellow Triangle
```

Save and close the Shapes program.

Using the `if-else-if-else` Statement: Updating the Thermometer Example

Let's return to the `Thermometer` program from Chapter 3. Choose **File** > **Save As** and resave this program as **TempColors**.

```
TempColors  ▼
int tempera
temperature
println("Te
```

Recall, when we run this program,

```
int temperature;
temperature = int( random(100) );
println("Temperature: " + temperature + char(176) + "F" );

size(150, 50);

fill(255, 0, 0);
rect(0, 15, temperature, 20);
fill(0, 0, 0);
text(temperature, temperature, 15);
```

an `int` value from 0 to 100 is chosen as an outdoor temperature and is output to the console:

Temperature: 60°F

On the canvas, a red rectangle is drawn that has a pixel length equal to this temperature (e.g., 60°, 60 pixels). This temperature is also displayed in black text at the end of the rectangle:

Suppose that we would like to have more color possibilities for our thermometer program so that the color of the rectangle drawn provides a color-coded general description of the temperature: blue (cold), light blue (cool), yellow (mild), orange (warm), and red (hot).

Temperature Range (°F):	Thermometer Color (Meaning) :
below 40	blue (cold)
40 to 49	light blue (cool)
50 to 64	yellow (mild)
65 to 79	orange (warm)
80 and above	red (hot)

An `if-else-if-else` statement will work very well for this. Notice that we have a *continuous* range of values:

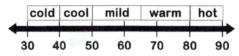

Whenever we would like to implement such a continuous range using an `if-else-if-else` statement, we need to identify the **breakpoints** in the range:

Temperature Range (°F):	Thermometer Color (Meaning) :
below 40	blue (cold)
40 to 49	light blue (cool)
50 to 64	yellow (mild)
65 to 79	orange (warm)
80 and above	red (hot)

breakpoints for: 40, 50, 65, 80

These breakpoints are the four temperatures that divide the range of temperatures into the five categories. For example, the breakpoint between the blue (cold) category and the light blue (cool) category occurs at 40°F. More specifically, the light blue (cool) category begins at 40°F, and any lower temperature falls into the blue (cold) category. Thus, we can test whether a temperature falls into the blue (cold) category and set the corresponding fill color by inserting the following:

```
if (temperature < 40)
{
  fill(0, 0, 255); // blue
  println("Cold");
}
```

The next breakpoint is 50°F, which divides the light blue (cool) category from the mild (yellow) category. Thus, we add the following `else if` clause:

```
if (temperature < 40)
{
  fill(0, 0, 255); // blue
  println("Cold");
}

else if (temperature < 50)
{
  fill(127, 226, 255); // light blue
  println("Cool");
}
```

At first glance, this logic might seem to be incorrect. After all, consider a temperature such as 32°F. This temperature is not only less than 40 but also less than 50. We might worry about the fact that 32°F not only satisfies the condition in our `if` clause,

```
if (temperature < 40)
{
  fill(0, 0, 255); // blue
  println("Cold");
}
```

but also satisfies the condition in the else if clause:

```
else if (temperature < 50)
{
  fill(127, 226, 255); // light blue
  println("Cool");
}
```

Thus, we might be concerned that the fill color for 32°F will be incorrectly set to light blue and incorrectly identified "cool." However, remember that if-else-if-else statements operate according to *short-circuit* evaluation. This means that as soon as a condition in one of the clauses of an if-else-if-else statement is determined to be true, then any *following* else if or else clauses are *not* performed. For example, if the temperature is 32°F, then the condition in the if clause is true. Thus, the fill color is set to blue, and this temperature is identified as "cold" in the console. Afterward, because of short-circuit evaluation, Processing skips over any else if or else clauses that follow. In this particular case, our else if clause testing if the temperature is less than 50°F is *not* performed:

```
if (temperature < 40)   ←true!
{
  fill(0, 0, 255); // blue
  println("Cold");
}

else if (temperature < 50)        ←not performed
{
  fill(127, 226, 255); // light blue
  println("Cool");
}
```

We can now complete our if-else-if-else statement by writing the following:

```
int temperature;
temperature = int( random(100) );
println("Temperature: " + temperature + char(176) + "F" );

size(150, 50);

if (temperature < 40)
{
  fill(0, 0, 255); // blue
  println("Cold");
}
```

```
else if (temperature < 50)
{
  fill(127, 226, 255); // light blue
  println("Cool");
}

else if (temperature < 65)
{
  fill(255, 255, 0); // yellow
  println("Mild");
}

else if (temperature < 80)
{
  fill(255, 127, 0); // orange
  println("Warm");
}

else // temperature is 80 or above
{
  fill(255, 0, 0); // red
  println("Hot");
}

rect(0, 15, temperature, 20);
fill(0, 0, 0);
text(temperature, temperature, 15);
```

Now, when we run the program, the correct temperature category is displayed in the console, and an appropriate colored rectangle is drawn on the canvas. These will be blue and "cold" (for temperatures less than 40°F),

```
Temperature: 29°F
Cold
```

or light blue and "cool" (for temperatures of 40°F–49°F),

```
Temperature: 47°F
Cool
```

or yellow and "mild" (for temperatures of 50°F–64°F),

```
Temperature: 52°F
Mild
```

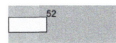

or orange and "warm" (for temperatures of 65°F–79°F),

```
Temperature: 71°F
Warm
```

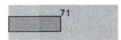

or red and "hot" (for temperatures of 80°F and above),

```
Temperature: 86°F
Hot
```

Short-circuit evaluation also means that whenever the condition of any given else if clause is performed, we can also assume that the conditions of any *preceding* clauses in the if statement have already proven to be *false*. For example, if the temperature is 71°F, then the conditions in the first three clauses of our if-else-if-else statement are performed, but they all turn out to be false. It is only after this that the condition for orange (warm) is performed and determined to be true. Thus, whenever the condition for orange (warm) is performed, we can assume that the conditions of three of the preceding clauses—for blue (cold), light blue (cool), and yellow (mild)—have *already* been performed and have all turned out to be *false*:

```
            false
if (temperature < 40)
{
   fill(0, 0, 255); // blue
   println("Cold");
}
            false
else if (temperature < 50)
{
   fill(127, 226, 255); // light blue
   println("Cool");
}
            false
else if (temperature < 65)
{
   fill(255, 255, 0); // yellow
   println("Mild");
}
```

```
                        true!
else if (temperature < 80)
{
  fill(255, 127, 0); // orange
  println("Warm");
}

else // temperature is 80 or above  ←not performed
{
  fill(255, 0, 0); // red
  println("Hot");
}
```

In other words, whenever we determine that the temperature is less than or equal to 80°F, we also *already* know that

- the temperature is *not* less than 40°F (the blue condition)

- the temperature is *not* less than 50°F (the light blue condition)

- the temperature is *not* less than 65°F (thc ycllow condition)

Thus, because of short-circuit evaluation, an if-else-if-else statement is able to implement continuous categorizing scales like the one in this example. We can visualize the logic of this if-else-if-else statement as the following flowchart:

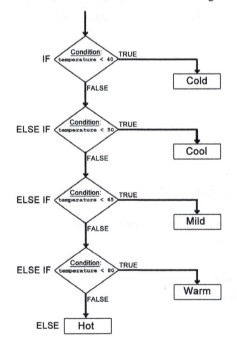

Save and close the TempColors program.

Using the if-else-if-else Statement: Types of Angles

Angles are sometimes classified according to the following five types:

acute   right   obtuse   straight   reflex

Type of Angle	Description
Acute	less than 90°
Right	90° exactly
Obtuse	greater than 90° but less than 180°
Straight	180° exactly
Reflex	greater than 180°

Let's write a program to draw and classify an angle as one of these types. Return to the DrawAngle program from Chapter 3. Select **File > Save As** and resave this program as **AngleTypes**.

AngleTypes ▼

import stat

Our current version of the program is

```
import static javax.swing.JOptionPane.*;

size(200, 200);

int degrees;
degrees = int( showInputDialog("Enter angle in degrees:") );
println("angle: " + degrees + char(176));

noFill();
strokeWeight(3);
arc(width / 2, height / 2, 150, 150, -radians(degrees), 0, PIE);

stroke(255, 255, 255);
arc(width / 2, height / 2, 150, 150, -radians(degrees), 0);
```

When we run this program, the user enters an angle measure in degrees:

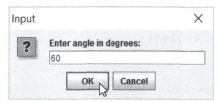

Input   ✕

?   Enter angle in degrees:
     60

          OK   Cancel

This entry is displayed back to the user in the console,

```
angle: 60°
```

and a representation of this angle is drawn on the canvas:

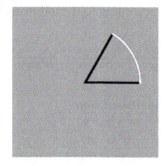

Now, let's add an `if-else-if-else` statement to identify the type of angle entered:

```
import static javax.swing.JOptionPane.*;

size(200, 200);

int degrees;
degrees = int( showInputDialog("Enter angle in degrees:") );
println("angle: " + degrees + char(176));

noFill();
strokeWeight(3);
arc(width / 2, height / 2, 150, 150, -radians(degrees), 0, PIE);

stroke(255, 255, 255);
arc(width / 2, height / 2, 150, 150, -radians(degrees), 0);

if (degrees < 90)
{
    println("acute angle");
}

else if (degrees == 90)
{
    println("right angle");
}

else if (degrees < 180)
{
    println("obtuse angle");
}

else if (degrees == 180)
{
    println("straight angle");
}
```

```
else // greater than 180
{
    println("reflex angle");
}
```

We can visualize the logic of this `if-else-if-else` statement as the following flowchart:

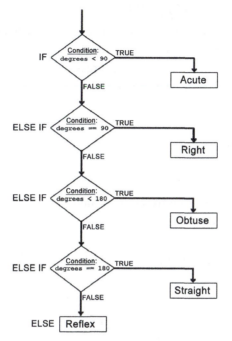

When we run this program, the user enters an angle:

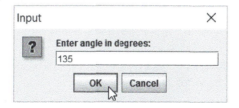

A representation of this angle is drawn upon the canvas:

However, now, not only the measure of the angle but also its type is displayed in the console:

```
angle: 135°
obtuse angle
```

Save and close the `AngleTypes` program.

Logical Operations: AND (&&), OR (||)

Recall that a *condition* is an example of what is known as a *Boolean* expression in that it always evaluates to either true or false.

In Processing, we can *combine* Boolean expressions together into what are known as **compound Boolean expressions**. We can do this using two *logical operators*:

1) `&&` The logical **AND** operator

2) `||` The logical **OR** operator

Using Logical AND (&&) : Rolling Two Dice

Given two conditions, we can require *both* of them to be true by joining them together using the `&&` operator, which represents a logical **AND** operation. As an example of the logical AND (`&&`) operation, let's modify our `Roll` program so that we can roll a *pair* of dice. Return to the `Roll` program. Select **File > Save As** and save this program as **RollDouble**.

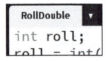

We'll start by renaming our `roll` variable to `roll1` and updating the console output accordingly:

```
int roll1;
roll1 = int( random(1, 7) );
println("Roll 1: " + roll1);

if (roll1 == 6)
{
  println("You won!");
}
else
{
  println("Sorry, you lost.");
}
```

We'll use copy and paste to duplicate the first three lines of code. Next, we'll rename the `roll1` variable in the duplicate lines of code to `roll2` and update the console output accordingly so that our program is now the following:

```
int roll1;
roll1 = int( random(1, 7) );
println("Roll 1: " + roll1);

int roll2;
roll2 = int( random(1, 7) );
println("Roll 2: " + roll2);
if (roll1 == 6)
{
  println("You won!");
}
else
{
  println("Sorry, you lost.");
}
```

Now, when we run this program, the rolls of *two* dice are output to the console:

```
Roll 1: 5
Roll 2: 4
```

Next, suppose we would like the player to win if the rolls of *both* dice are 6. For this, we can use two conditions joined by a logical AND operation:

(roll1 == 6) && (roll2 == 6)

Thus, our program can be modified to

```
int roll1;
roll1 = int( random(1, 7) );
println("Roll 1:", roll1);

int roll2;
roll2 = int( random(1, 7) );
println("Roll 2:", roll2);

if ( (roll1 == 6) && (roll2 == 6) )
{
  println("You won!");
}
else
{
  println("Sorry, you lost.");
}
```

When we run the program and *neither* of the dice is a 6, the player loses. For example,

```
Roll 1: 5
Roll 2: 4
Sorry, you lost.
```

When we run the program and only *one* of the dice is a 6, the player loses. Examples

```
Roll 1: 2
Roll 2: 6
Sorry, you lost.
```

```
Roll 1: 6
Roll 2: 3
Sorry, you lost.
```

However, when *both* dice are 6, the player wins:

```
Roll 1: 6
Roll 2: 6
You won!
```

In a logical AND (&&) operation, *both* conditions must be true in order for the overall compound Boolean expression to be true.

Save the RollDouble program. We will return to it later in this chapter.

Using Logical OR (||) : Rolling a Die

We can also create a compound Boolean expression by joining two conditions together using the logical **OR** operation represented by the "||" operator. When we do this, we require that *at least one* of the two conditions must be true. Let's work through an example of a logical OR operation. Return to the Roll program. Select **File > Save As** and save this program as **RollOneOrSix**.

```
RollOneOrSix  ▾
int roll;
roll = int(
```

This program simulates the roll of a single die. Currently, the player wins only when a 6 is rolled. However, what if we wish to have the player win if a 1 *or* a 6 is rolled?

For this, we can write the following:

```
int roll;
roll = int( random(1, 7) );
println("Roll: " + roll);

if ( (roll == 1) || (roll == 6) )
{
  println("You won!");
}
else
{
  println("Sorry, you lost.");
}
```

Now, the player wins if a 1 is rolled:

```
Roll: 1
You won!
```

But the player also wins if a 6 is rolled:

```
Roll: 6
You won!
```

Otherwise, the player loses, for example,

```
Roll: 5
Sorry, you lost.
```

Save and close this RollOneOrSix program. We will return to it later in this chapter.

Logical OR (||) Example: Rolling Two Dice

Here's another example of using a logical OR operation. Return to the RollDouble program. Select **File > Save As** and save this program as **RollSix**.

In the current version of this program, a player wins only if *both* dice rolls are 6:

```
int roll1;
roll1 = int( random(1, 7) );
println("Roll 1:", roll1);

int roll2;
roll2 = int( random(1, 7) );
println("Roll 2:", roll2);
if ( (roll1 == 6) && (roll2 == 6) )
{
  println("You won!");
}
else
{
  println("Sorry, you lost.");
}
```

Suppose that we would instead like to have the player win if a 6 is rolled on *either* of the two dice.

For this, we can join the two conditions with a logical OR (||):

```
int roll1;
roll1 = int( random(1, 7) );
println("Roll 1:", roll1);

int roll2;
roll2 = int( random(1, 7) );
println("Roll 2:", roll2);
if ( (roll1 == 6) || (roll2 == 6) )
{
  println("You won!");
}
else
{
  println("Sorry, you lost.");
}
```

Now, when we run the program and *neither* of the dice is a 6, the player *loses*. For example,

```
Roll 1: 5
Roll 2: 4
Sorry, you lost.
```

However, the player *wins* when the roll of the *first* die is a 6. For example,

```
Roll 1: 6
Roll 2: 3
You won!
```

Likewise, the player *wins* when the roll of the *second* die is a 6. For example,

```
Roll 1: 2
Roll 2: 6
You won!
```

However, notice that the player also *wins* when *both* dice are 6:

```
Roll 1: 6
Roll 2: 6
You won!
```

This last roll illustrates how a logical OR operation also evaluates to true when *both* conditions are true. A logical OR operation does not require *only one* of the conditions to be true.* In other words, the logical OR operation is *not* paraphrased as

"one or the other and *not both*"   ←**incorrect**

Rather, a logical OR operation requires *at least one* of the conditions to be true. Thus, the logical OR operation can be paraphrased as

"one or the other *or both*"   ←**correct**

Save and close the RollSix program.

Nested `if` Statements

Sometimes, when we ask a question, the answer we receive compels us to ask a follow-up question. Similarly, another way to implement *multiple* conditions in Processing is by *nesting* an if statement inside another if statement. The simplest form of nesting if statements is

---

* The logical OR operation implemented by Processing's || operator is *not* an *exclusive* OR logical operation. (The exclusive OR logical operation is often written as "XOR.")

```
if (condition-1)
{
  if (condition-2)
  {
      Statements to perform if both condition-1 and condition-2 are true
  }
}
```

Using Nested if Statements: Classifying Ages

Let's try an example that will illustrate nested if statements. Return to the Ages program. Select **File > Save As** and resave this program as **NestedAge**.

Let's expand the age range in our program to be from 1 up to (but not including) 100. Thus, our program is now

```
int age;
age = int( random(1, 100) );
println("Age: " + age);

if (age >= 18)
{
  println("Adult");
}

else
{
  println("Minor");
}
```

When we run this program, we see from the console output that the age chosen is now classified as either the age of an adult,

```
Age: 52
Adult
```

or the age of a minor,

```
Age: 14
Minor
```

However, suppose that if the age is that of an adult, we would like to identify also whether this is the age of someone who is 65 or older, an age commonly described as a "senior." This is comparable to asking the follow-up question,

"Is the person also aged 65 or older?"

To modify our program to ask this follow-up question, we need to add an `if` statement *inside* our existing `if` clause:

```
int age;
age = int( random(1, 100) );
println("Age: " + age);

if (age >= 18)
{
  println("Adult");

  if (age >= 65)
  {
    println("Senior");
  }
}
else
{
  println("Minor");
}
```

This is an example of a *nested* `if` statement. The *outside* `if` statement is sometimes known as the **outer** `if` statement. The *inside* `if` statement is sometimes known as the **inner** `if` statement.

```
outer         ┌──→ if (age >= 18)
if            │    {
statement     │        println("Adult");
              │
              │        if (age >= 65)    ←──┐ inner
              │        {                    │ if
              │            println("Senior"); │ statement
              │        }                 ←──┘
              └──→ }
```

Now, when we run this program, the age of a minor is identified:

```
Age: 6
Minor
```

And the age of a nonsenior adult is identified as before:

```
Age: 23
Adult
```

However, whenever the age is 65 or older, it is identified not only as the age of an adult but also as the age of a senior:

```
Age: 72
Adult
Senior
```

Similarly, if we would like to check whether a given age of a minor is also the age of a preteen, we can nest the following `if` statement inside our `else` clause:

```
int age;
age = int( random(1, 100) );
println("Age: " + age);

if (age >= 18)
{
  println("Adult");

  if (age >= 65)
  {
    println("Senior");
  }
}

else
{
  println("minor");

  if (age < 13)
  {
    println("Preteen");
  }
}
```

Now, when we run our program, an age identified as that of a minor will also be evaluated to determine whether it is the age of a preteen. Thus, in the case of a minor who is a teen, we see output like before:

```
Age: 14
Minor
```

However, we see that a preteen minor is now also identified as such:

```
Age: 5
Minor
Preteen
```

As we have seen, a nested `if` statement essentially enables us to pose a follow-up question inside an `if`, `else if`, or `else` clause. Nested `if` statements won't be used often in this book, but it is good to have a basic understanding of them.

Save and close the `NestedAge` program.

boolean Variables

As we have learned, a *Boolean* expression always turns out to be either *true* or *false*. Each of the *conditions* we have used thus far in our `if` statements is an example of such a Boolean expression.

In Processing, we are also allowed to define *variables* that are of the **boolean** type. For example, suppose that we were writing a program to ask the user a true-or-false question. We could create a variable to store the correct answer to this true-or-false question by writing

```
boolean correctAnswer;
```

A `boolean` variable is only able to store one of the two possible `boolean` values: **true** or **false**. Suppose the question that we want to ask the user is,

"TRUE or FALSE? A tomato is a fruit."

The correct answer to this true-and-false question is *true*. Thus, a program to ask the user this question might include the following:

```
boolean correctAnswer;
correctAnswer = true;
```

Or, suppose the question that we want to ask the user is,

"TRUE or FALSE? A peanut is a nut."

The correct answer to this true-and-false question is *false*. Thus, a program to ask the user this question might include the following statements:

```
boolean correctAnswer;
correctAnswer = false;
```

Using a `boolean` Variable to Store a Condition

One common use of a `boolean` variable is to store the result of a *condition*. Let's work through an example to illustrate this.

Return to the `Ages` program. Select **File** > **Save As** and resave this program as `BooleanVariable`.

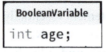

BooleanVariable

```
int age;
```

Our program currently determines whether a randomly chosen age is that of an adult (18 or older) or a minor (under 18):

```
int age;
age = int( random(1, 30) );
println("Age: " + age);

if (age >= 18)
{
  println("Adult");
}
else
{
  println("Minor");
}
```

When we run this program, we see that it properly detects the age of a minor,

```
Age: 14
Minor
```

and the age of an adult,

```
Age: 22
Adult
```

If we like, we can declare a `boolean` variable named `adult`:

```
boolean adult;
```

We can visualize this variable declaration as

boolean

adult

A `boolean` variable can store only one of the two `boolean` values: `true` or `false`. We can visualize this as

boolean **true**

adult

or

$$\boxed{\text{false}}$$
boolean | false
adult

Note that boolean values are *not* String values (*not* "true" and *not* "false").

Once a boolean variable is declared, it can be assigned a boolean value using an assignment statement such as

```
boolean adult;
adult = true;
```

or

```
boolean adult;
adult = false;
```

A boolean variable can also be assigned the result of a comparison as we might find in a condition. For example, we could write

```
boolean adult;
if (age >= 18)
{
  adult = true;
}
else
{
  adult = false;
}
```

However, we can accomplish the same result by simply writing

```
boolean adult;

adult = (age >= 18);
```

At first glance, the above assignment statement is somewhat unusual, so let's take a closer look at it. Remember, whenever there is an *expression* on the *right*-hand side of an assignment statement, Processing must evaluate this expression *first*, before the assignment operation can be performed. In this particular case, the right-hand side contains a boolean expression:

$$\text{adult} = \boxed{(\text{age} >= 18)};$$

Thus, this boolean expression on the right-hand side must be evaluated *first*, before the assignment operation can be performed. A boolean expression always evaluates to one of the two possible boolean values: either true or false. For example, suppose the value of age is 23. In this case, the `boolean` expression evaluates to the `boolean` value `true`. We can visualize this evaluation as

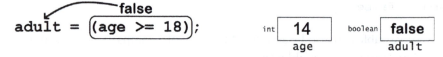

As a result, the `boolean` value `true` is assigned to the `boolean` variable `adult`. We can visualize this assignment operation and its result as

On the other hand, suppose the value of `age` is 14. In this case, the `boolean` value `false` is assigned to `adult`. We can visualize this assignment operation as

To verify this result, we can print out the value of `adult` to the console:

```
boolean adult;
adult = (age >= 18);
println("adult is: " + adult);
```

Our program is now

```
int age;
age = int( random(1, 30) );
println("Age: " + age);

boolean adult;
adult = (age >= 18);
println("adult is: " + adult);

if (age >= 18)
{
  println("Adult");
}
else
{
  println("Minor");
}
```

When we run our program, we see that, depending on the value of the `age` variable, our `boolean` variable `adult` is indeed being assigned either the `boolean` value `true`,

```
Age: 20
adult is: true
Adult
```

or the `boolean` value `false`,

```
Age: 12
Adult is: false
Minor
```

Our `boolean` variable `adult` is being assigned either `true` or `false` based on the result of the same condition that we are using in our `if-else` statement:

```
int age;
age = int( random(1, 30) );
println("Age: " + age);

boolean adult;
adult = (age >= 18);
println("adult is: " + adult);

if (age >= 18)
{
  println("Adult");
}
else
{
  println("Minor");
}
```

For this reason, we can actually use our `boolean` variable `adult` in our `if-else` statement, *in place of* the condition that is currently there. Thus, we can modify our `if-else` statement to

```
if (adult == true)
{
  println("Adult");
}
else
{
  println("Minor");
}
```

or, simply,

```
if (adult)
{
  println("Adult");
}
else
{
  println("Minor");
}
```

Thus, our program is now

```
int age;
age = int( random(1, 30) );
println("Age: " + age);

boolean adult;
adult = (age >= 18);
println("adult is: " + adult);

if (adult)
{
  println("Adult");
}
else
{
  println("Minor");
}
```

When we run this modified version of the program, we see that the age of a minor is still correctly identified:

```
Age: 16
adult is: false
Minor
```

And the age of an adult is still correctly identified as well:

```
Age: 23
adult is: true
Adult
```

Thus, as we have seen, a boolean variable can be used to *store* a condition:

```
adult = (age >= 18);
```

and a boolean variable can be used *as* a condition*,

```
if (adult == true)
```

or

```
if (adult)
```

---

* One reason for using the second form of a boolean variable within a condition
  if (adult)
is because an error that can be difficult to find can result if the second form
  if (adult == true)
is used but is mistakenly typed using a *single* equals sign:
  if (adult = true)

It takes time to become accustomed to working with `boolean` variables. Rest assured, they aren't used often in this book. For now, just remember that a `boolean` variable is a handy way to store the result of a condition.

The `switch` Statement

When we are testing for individual values of an `int`, `char`, or `String` variable in an `if-else-if-else` statement, we also have the option of using another type of selection structure that is known as the **switch** statement. The basic form of a `switch` statement is difficult to describe in general terms. It is easier to learn the syntax of the `switch` statement through examples. Thus, the next several sections feature examples that make use of the `switch` statement.

Using the `switch` Statement: Revisiting the Rock–Paper–Scissors Example

Let's return to our `Rock` program. Select **File** > **Save As** and resave this program as **RockSwitch**.

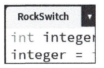

Recall, when we run this program, it randomly selects an integer—1, 2, or 3—and then displays "rock," "paper," or "scissors" in the console, respectively. Currently, this program uses an `if-else-if-else` statement:

```
int integer;
integer = int( random(1, 4) );
println("integer is: " + integer);

if (integer == 1)
{
  println("Computer picks: Rock!");
}

else if (integer == 2)
{
  println("Computer picks: Paper!");
}

else // integer is 3
{
  println("Computer picks: Scissors!");
}
```

When we test for individual values of an `int` or `char` variable in an `if-else-if-else` structure like this, we have the option of using a `switch` statement instead.

The overall structure of the **switch** statement consists of the keyword `switch`, followed by the variable containing the value we are checking:

```
switch (variable)
{

}
```

Thus, in our current rock–paper–scissors example, we begin with

```
int integer;
integer = int( random(1, 4) );
println("integer is: " + integer);

switch (integer)
{

}
```

Within this overall structure, we then specify each "case" of the variable that we wish to test. Each of these cases consists of the keyword **case**, followed by a specific value of the variable and a colon. We then specify the statements that we wish to have executed when the value of this variable is equal to this particular case's value. Thus, in place of

```
if ( integer == 1 )
{
  println("Computer picks: Rock!");
}
```

we write

```
switch (integer)
{
 case 1 :
 {
   println("Computer picks: Rock!");
   break;
 }
}
```

Notice that we conclude this case by adding a **break** statement. This is to ensure that Processing does not go on to the next case, but rather exits this `switch` statement.

We now can construct our second case, where we check if `integer` is equal to 2. Thus, in place of

```
else if ( integer == 2 )
{
  println("Paper!");
}
```

we add

```
switch (integer)
{
  case 1 :
  {
    println("Computer picks: Rock!");
    break;
  }

  case 2 :
  {
    println("Computer picks: Paper!");
    break;
  }
}
```

Lastly, similar to the `else` clause in our `if` statement, we can specify an optional `default` case that will be performed in the event that *none* of the preceding cases is true. Thus, in place of

```
else
{
  println("Scissors!");
}
```

we add

```
switch (integer)
{
  case 1 :
  {
    println("Computer picks: Rock!");
    break;
  }
```

```
case 2 :
{
    println("Computer picks: Paper!");
    break;
}
default :
{
    println("Computer picks: Scissors!");
}
}
```

Thus, our program is now

```
int integer;
integer = int( random(1, 4) );
println("integer is: " + integer);

switch (integer)
{
 case 1 :
 {
    println("Computer picks: Rock!");
    break;
 }

 case 2 :
 {
    println("Computer picks: Paper!");
    break;
 }

 default :
 {
    println("Computer picks: Scissors!");
 }
}
```

When we run this program, we see the same results using a `switch` statement as we did using an `if-else-if-else` statement:

```
integer is: 1
Computer picks: Rock!
```

or

```
integer is: 2
Computer picks: Paper!
```

or

```
integer is: 3
Computer picks: Scissors!
```

Save and close the RockSwitch program.

Combining switch Cases: Rolling a One or a Six

Let's return again to our RollOneOrSix program. Select **File** > **Save As** and resave this program as **RollSwitch**.

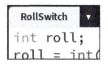

The logical OR operation in our program currently specifies that the player wins if a 1 or a 6 is rolled:

```
int roll;
roll = int( random(1, 7) );
println("Roll: " + roll);

if ( (roll == 1) || (roll == 6) )
{
  println("You won!");
}
else
{
  println("Sorry, you lost.");
}
```

We can modify this program to use a switch statement by writing

```
int roll;
roll = int( random(1, 7) );
println("Roll: " + roll);

switch (roll)
{
 case 1:
 {
   println("You won!");
   break;
 }
```

```
  case 6:
  {
    println("You won!");
    break;
  }

  default:
  {
    println("Sorry, you lost.");
  }
}
```

However, the `switch` statement actually allows us to *combine* cases. Thus, we can simplify this program to the following:

```
int roll;
roll = int( random(1, 7) );
println("Roll: " + roll);

switch (roll)
{
 case 1:
 case 6:
 {
   println("You won!");
   break;
 }

 default:
 {
   println("Sorry, you lost.");
 }

}
```

When we run this program, we see the same results as before, where the player wins whenever a 1 or a 6 is rolled.

Save and close the `RollSwitch` program.

Another `switch` Statement Example: Days of the Month

Let's write a program that will identify the number of days in a month. A `switch` statement will work nicely for this. Select **File > New** to start a new program. Save this program as **MonthDays**.

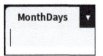

We'll start by reading in a month number from the user and storing it in a `String` variable named `input`:

```
import static javax.swing.JOptionPane.*;

String input;
input = showInputDialog("Enter month number:");
```

Remember that when we use an input dialog, the input from the user returned to us by the `showInputDialog()` function is always a `String` value, even when the user enters a numerical value. Thus, in order to convert the user's input from a `String` value back into the corresponding numerical value, we must use the `int()` or `float()` function to do so. Here, we'll use the `int()` function for this purpose:

```
import static javax.swing.JOptionPane.*;

String input;
input = showInputDialog("Enter month number:");

int month;
month = int(input);
```

Next, we'll use an `if-else-if-else` statement to determine the number of days, given the month number, and then output that number of days in the console:

```
import static javax.swing.JOptionPane.*;

String input;
input = showInputDialog("Enter month number:");

int month;
month = int(input);

if (month == 2)
{
  println("That month has 28 or 29 days");
}
else if ( (month == 4) || (month == 6) ||
          (month == 9) || (month == 11) )
{
  println("That month has 30 days.");
}
else
{
  println("That month has 31 days.");
}
```

When we run this program, the user enters a month number,

```
Input                                    ✕

  ?    Enter month number:
       6

            OK      Cancel
```

and the corresponding number of days is displayed:

That month has 30 days.

As we have learned, the switch statement allows us to combine cases that have the same corresponding actions. Thus, here is a version of the above program that instead uses the switch statement:

```
int month;
month = int(input);

switch (month)
{
  case 2:
  {
    println("That month has 28 or 29 days.");
    break;
  }

  case 4:
  case 6:
  case 9:
  case 11:
  {
    println("That month has 30 days.");
    break;
  }

  default:
  {
    println("That month has 31 days.");
  }
}
```

When we run this version of the program, we see the same kind of console output as that produced by the if-else-if-else version.

Save and close the MonthDays program.

Checking Equality: `float` Values

Because of the imprecision with which computers store fractions, we have to be careful about testing for equality of **float** values. For example, consider the following program:

```
float first;
first = 1.0;

float second;
second = 0.99999999;

if (first == second)
{
    println("Equal!");
}
```

If we run this program, the output to the console is

Equal!

However, we can see that the `float` values stored in these two variables are *not* equal.

Whenever possible, use greater than (>), greater than or equal to (>=), less than (<), or less than or equal to (<=) operators when comparing `float` values.

Checking Equality: `String` Objects

Testing whether two `String` values are equal requires us to adopt special measures.

Select **File** > **New** to start a new program. Save this program as **StringEquals**.

```
StringEquals  ▼
|
```

Enter the following statements into the program:

```
import static javax.swing.JOptionPane.*;

String day;
day = showInputDialog("Enter the first day of the week:");

println("You entered: " + day);
```

When we run this program and enter a day of the week,

```
Input                                    ✕

  ?    Enter the first day of the week:
       Sunday

            OK       Cancel
```

the day of the week that we entered is displayed back to us in the console:

```
You entered: Sunday
```

Now, how do we determine whether the user has correctly entered "Sunday," the first day of the week? An `if-else` statement will work nicely for this. It is tempting to use the equals comparison operator consisting of two equals signs (`==`) to compare two `String` objects. Thus, we might try inserting the following `if-else` statement:

```
import static javax.swing.JOptionPane.*;

String day;
day = showInputDialog("Enter the first day of the week:");

println("You entered: " + day);
if (day == "Sunday")
{
  println("That is the first day of the week.");
}
else
{
  println("That is not the first day of the week.");
}
```

However, when we run this program, we find that no matter what, even when we enter "Sunday,"

we are told that this is not the first day of the week:

```
You entered: Sunday
That is not the first day of the week.
```

The problem we are encountering here is that the condition in our `if` clause,

```
if (day == "Sunday")
```

is *never* evaluating to `true`. Rather, it is evaluating to `false`, no matter what day we enter. Thus, every time we run the program, the `else` clause of our `if-else` statement is always performed:

```
if (day == "Sunday")
{
  println("That is the first day of the week.");
}
else
{
  println("That is not the first day of the week.");
}
```

Why is this happening? The int, float, char, and boolean types are all *primitive* types, the most basic types of values and variables in Processing. A variable of a primitive type is a named storage location that actually contains whatever primitive value is assigned to it. For example, when we write statements like

```
int first;
first = 15;
```

we can visualize the result as

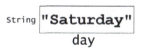

String is an **object** type, a slightly more complex type of value and variable in Processing. When we write statements like

```
String day;
day = "Saturday";
```

we can visualize the result as

String "**Saturday**"
day

However, behind the scenes, a String variable does not contain an actual sequence of characters. Rather, it contains the *address* in the computer's memory where that sequence of characters is stored. Thus, a more accurate visualization might be something like

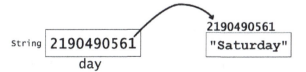

Thus, behind the scenes, whenever we compare two items of type String using the equals comparison operator (==), Processing actually compares the two memory *addresses*, not the sequence of characters that are stored at those locations.

The details concerning the way String objects are handled behind the scenes in Processing are rather complicated. What's most important to be aware of, here, is that this comparison of String values and variables using == will only *sometimes* yield a correct result (`true` or `false`), so it should *never* be used with `String` variables and values. Therefore, we need to use another technique whenever we compare `String` values and variables to see if they are equal, a technique that we will learn in the next section.

Save the `StringEquals` program. We will continue working with it in the next section.

The `equals()` Method of a `String` Object

As we have learned, `String` is actually an *object* type. One of the features of an object is that it can have predefined *functions* built into it. Such a function that is built into an object is also known as a "**method**." (Thus, whenever you hear the word *method*, you may simply think of it as a predefined function that is built into an object.) For example, every `String` object has a built-in **`equals()`** method for *reliably* checking if that `String` object has the same `String` value as another `String` object. This `equals()` method takes one argument: the other `String` object to which we are comparing the current `String` object. To use this method, we simply add a period (dot) to the name of the `String` variable, followed by the call to the `equals()` method. Calling the `equals()` method in this manner is an example of what is known as **dot notation** or **dot syntax**. Return to the `StringEquals` program. Here's how we can reliably check if the `String` variable named day is equal to the `String` value `"Sunday"`:

```
day.equals("Sunday")
```

Thus, the basic form of testing for equality using the `equals()` method is

```
String1.equals(String2)
```

Once we make the appropriate modifications to our program, we now have the following:

```
import static javax.swing.JOptionPane.*;

String day;
day = showInputDialog("Enter the first day of the week:");

println("You entered: " + day);
if ( day.equals("Sunday") )
{
  println("That is the first day of the week.");
}

else
{
  println("That is not the first day of the week.");
}
```

Now, when we run this program, we find that when we enter "Sunday,"

it is properly identified as the first day of the week:

```
You entered: Sunday
That is the first day of the week.
```

Save and close this program.

Summary

In this chapter, we learned about the powerful principle of **selection**, which enables us to specify actions that we want to be performed when certain **conditions** are true.

- We learned about the basic `if` statement, as well as its `if-else` and `if-else-if-else` forms.

- We learned about `boolean` *expressions*, such as those used in the *condition* of an `if` statement, and how to create them using the comparison (relational) operators: $<$, $>$, $<=$, $>=$, $==$, and $!=$.

- We learned how to create *compound* Boolean expressions using the AND (`&&`) and OR (`||`) operations.

- We learned about `boolean` *variables* and how they can be used to store conditions.

- We learned about the `switch` statement, which can take the place of `if-else-if-else` structures when mutually exclusive conditions are required.

- We learned about reliably testing for equality of `String` objects using the `equals()` method.

Exercises

Using if

1) Write a program to do the following:

 a) Declare a variable named age that can store an integer.

 b) Assign age a random integer from 1 up to (and not including) 50.

 c) Output to the console "age:", followed by the value of age.

 d) If the value of age is 18 or greater, display "adult" in the console.

 e) Run your program several times to make sure that it works correctly for both adult and nonadult ages.

 Sample runs

    age: 11   age: 21

                   adult

Using if-else

2) Modify the previous program by adding an else clause so that "minor" is displayed when the age is less than 18. Run your program several times to make sure that it works for both adult and minor ages.

 Sample runs

    age: 43   age: 6

    adult   minor

3) Write a program to do the following:

 a) Declare a variable named testScore that can hold a number with decimal places.

 b) Assign testScore a random number between 0 and 100.

 c) Output to the console "Test score:", followed by the value of testScore.

 d) If the value of testScore is greater than or equal to 65, output "Pass" to the console.

 e) Otherwise, output "Fail" to the console.

4) Write a program to do the following:

   a) Declare a variable named `toss` that can store a number with decimal places.

   b) Assign `toss` a random number from 0 up to (and not including) 2.

   c) Display "toss:" in the console, followed by the value of toss.

   d) If the value of toss is greater than or equal to 1, display "Heads!" to the console.

   e) Otherwise, display "Tails!" to the console.

5) Write a program to do the following:

   a) Set the canvas size to 100 pixels by 200 pixels.

   b) Declare a variable named `row` and assign it a random integer from 0 to the height of the canvas.

   c) Display "row:", followed by the value of `row` in the console.

   d) Using an `if-else` statement, set the stroke color to green if the selected row number is less than 100; otherwise, set the stroke color to blue.

   e) Using the `row` and `width` variables, draw a horizontal line segment across the entire width of the canvas at the chosen row number:

   ```
   line(0, row, width, row);
   ```

6) Write a program to do the following:

   a) Declare a variable named `integer` that can hold an integer.

   b) Assign `integer` a random integer from 1 up to (but not including) 100.

   c) Display the value of `integer` in the console.

   d) Declare a variable named `remainder` that can hold an integer.

   e) Use the modulo (`%`) operator to divide `integer` by 2 and store the result in `remainder`.

   f) Display to the console

   ```
   "___ divided by 2 gives remainder: ___"
   ```

   Use the `integer` and `remainder` variables to fill in the blanks.

   g) If the value of `remainder` is 0, output to the console

   ```
   "____ is an even number."
   ```

   Use the `integer` variable to fill in the blank.

   h) If the value of `remainder` is 1, output to the console

   ```
   "____ is an odd number."
   ```

   Use the `integer` variable to fill in the blank.

Using if-else-if-else

7) Write a program to implement the grading scale shown below:

90 - 100:	A
80 - 89:	B
70 - 79:	C
60 - 69:	D
0 – 59:	F

a) Declare a variable named score that can hold an integer.

b) Assign score a random integer from 0 up to (and including) 100.

c) Output to the console "Score:", followed by the value of score.

d) Display "Grade:" to the console without advancing to the next line.

e) If the value of score is greater than or equal to 90, display "A" in the console.

f) Otherwise, if the value of score is greater than or equal to 80, display "B" in the console.

g) Otherwise, if the value of score is greater than or equal to 70, display "C" in the console.

h) Otherwise, if the value of score is greater than or equal to 60, display "D" in the console.

i) Otherwise, display "F" in the console.

j) Run your program several times and check the results.

Sample runs

```
Score: 96        Score: 84        Score: 62

Grade: A         Grade: B         Grade: D
```

8) Modify the previous program so that it instead does the following:

a) If the value of score is less than 60, display "F" in the console.

b) Otherwise, if the value of score is less than 70, display "D" in the console.

c) Otherwise, if the value of score is less than 80, display "C" in the console.

d) Otherwise, if the value of score is less than 90, display "B" in the console.

e) Otherwise, display "A" in the console.

f) Run your program several times and check the results.

g) Notice how this program produces the same results as the previous program using different conditions.

Sample runs

```
Score: 92        Score: 74        Score: 52

Grade: A         Grade: C         Grade: F
```

9) Write a program to do the following:

   a) Set the canvas size to 100 pixels by 400 pixels.

   b) Declare a variable named `row` and assign it a random integer from 0 to the height of the canvas.

   c) Display "row:", followed by the value of `row` in the console.

   d) Using an `if-else  if-else` statement, set the stroke color to green if the selected row number is less than 100, to blue if the selected row number is less than 200, and to yellow if the selected row number is less 300. Otherwise, set the stroke color to red.

   e) Using the `row` and `width` variables, draw a horizontal line segment across the entire width of the canvas at the chosen row number:

   ```
   line(0, row, width, row);
   ```

10) Write a program to implement the wind classification scale shown in the following table:

Category	Wind Speed
5	157 mph and above
4	130 - 156 mph
3	111 - 129 mph
2	96 - 110 mph
1	74 - 95 mph
Not a hurricane	73 mph and below

   a) Declare a variable named `windSpeed` that can store an integer.

   b) Assign to `windSpeed` a random integer from 50 up to (and including) 175.

   c) Display the value of `windSpeed` followed by "miles per hour" in the console.

   d) Display "category:" in the console without advancing to the next line.

   e) Using an `if-else  if-else` statement, display the hurricane category that corresponds to the value of `windSpeed`.

   f) Run your program several times and check the results.

   Sample runs

   ```
   64 miles per hour              138 miles per hour
   category: not a hurricane      category: 4
   ```

11) Write a short program that uses an `if-else` `if-else` statement to do the following:

    a) Pick a random integer between 1 and 5.

    b) Display this integer in the console.

    c) Display the corresponding name of this integer—that is, "one," "two," "three," "four," or "five"—in the console.

Using a `switch` Statement

12) Write a short program that produces the same results as the program in the preceding exercise but uses a `switch` statement instead.

Using Logical AND and OR

13) Write a program to do the following:

    a) Set the canvas size to 100 pixels by 300 pixels.

    b) Declare a variable named `row` and assign it a random integer from 0 to the height of the canvas.

    c) Display "row:", followed by the value of `row` in the console.

    d) Use an `if-else` statement that contains a logical AND operation (`&&`) to set the stroke color to green if the selected row number is greater than 100 and less than 300. Otherwise, set the stroke color to blue.

    e) Using the `row` and `width` variables, draw a horizontal line segment across the entire width of the canvas at the chosen row number:

```
line(0, row, width, row);
```

14) Write a program to do the following:

    a) Set the canvas size to 100 pixels by 300 pixels.

    b) Declare a variable named `row` and assign it a random integer from 0 to the height of the canvas.

    c) Display "row:", followed by the value of `row` in the console.

    d) Use an `if-else` statement with a condition containing a logical OR operation (`||`) to set the stroke color to green if the selected row number is less than 100 or greater than 300. Otherwise, set the stroke color to blue.

    e) Using the `row` and `width` variables, draw a horizontal line segment across the entire width of the canvas at the chosen row number:

```
line(0, row, width, row);
```

15) Write a program to do the following:

a) Pick a random integer that falls between 1 and 30.

b) Assign this integer to a variable named `age`.

c) Display the value of `age` in the console.

d) Use an `if-else` statement containing a logical AND test if the value of `age` is greater than or equal to 13 and less than or equal to 19, and if so, display "Teen" in the console. Otherwise, display "Not a teen" in the console.

Sample runs

```
Age: 13        Age: 6          Age: 21
Teen           Not a teen      Not a teen
```

16) Write a program to do the following:

a) Pick a random integer that falls between 1 and 30.

b) Assign this integer to a variable named `age`.

c) Display the value of `age` in the console.

d) Use an `if-else` statement containing a logical OR to test if the value of `age` is less than 13 or greater than 19, and if so, display "Not a teen" in the console. Otherwise, display "Teen" in the console.

Sample runs

```
Age: 15        Age: 11         Age: 20
Teen           Not a teen      Not a teen
```

Using a boolean Variable

17) Write a short program to do the following:

a) Pick a random integer that falls between 1 and 30.

b) Assign this integer to a variable named `age`.

c) Display the value of `age` in the console.

d) Declare a `boolean` variable named `teen`.

e) Assign `teen` the value `true` or `false` depending on whether or not `age` is greater than 12 and less than 20.

f) Use `teen` in the condition of an `if-else` statement to determine whether to display "Teen" or "Not a teen" in the console.

Sample runs

```
Age: 16    Age: 21
Teen       Not a teen
```

CHAPTER 5

Repetition with a Loop: The `while` Statement

We have seen in the programs we have written that the order in which the statements are listed makes a difference. This illustrates a key principle of computing known as **sequence**.

In Chapter 4, we saw that statements placed inside an `if` or `switch` statement are only performed if a certain condition is true. Such conditional programming illustrates a second key principle of computing known as **selection**.

In this chapter and the next, we will see that, sometimes, one or more statements need to be performed *repeatedly* in a computer program. This will illustrate a third key principle in computing: **repetition**.

Human beings often find repetitive tasks to be tedious. In contrast, computers can provide an ideal way to perform repetitive tasks.

Repetition as Long as a Certain Condition Is True

Like the `if` statement, each repetition structure in this chapter will make use of a *condition*. This condition will determine whether the repetition should *continue*.

Once again, let's consider a cooking recipe as an analogy to a computer program. Suppose a certain dough recipe says that you are to mix the ingredients and then *repeatedly* add ½ cup of flour *as long as* you can still stir the mixture with a spoon. A flowchart of this recipe might look like the following:

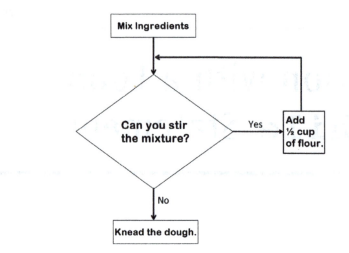

Similarly, in computer programming, there are sometimes sets of statements that we would like to perform *repeatedly*, but only *as long as* a certain condition is true.

Repetition with the `while` Statement

Processing provides several programming structures that can be used for creating repetition. One of the more versatile of these repetition structures is the **while** statement. The basic form of the while statement is the following:

```
while (condition)
{
    Statement(s) to continue to perform
    as long as condition remains true
}
```

where

> **condition** is the condition under which the statement(s) will continue to be performed, repeatedly.*

Because the while statement creates repetition, it is also known as a while **loop**. In this chapter, we will work through several examples that illustrate the functionality of a while statement (loop).

Using the `while` Statement: Rolling a Die Repeatedly

Let's return to our Roll program from Chapter 3. Select **File > Save As** and resave this program as **RollWhile**.

---

* Once again, if there is only one statement to be repeated, then technically, the pair of curly braces is unnecessary. However, we will always use curly braces in the while statements in this book. This is not only for consistency but also because most while loops have more than one statement to be repeated.

Currently, our program consists of the following:

```
int roll;
roll = int( random(1, 7) );
println("Roll: " + roll);
```

When we run this program, we see console output like the following:

Roll: 2

If we keep running the program, it is only a matter of time before we see that a 6 is rolled:

Roll: 6

We can be certain of this because we know that the statement

roll = int(random(1, 7));

will eventually assign the roll variable the value 6 if we repeatedly run the program.

Currently, our program simulates just a single roll of a die. Thus, if we wish to simulate *multiple* rolls of a die, we currently need to rerun the entire program. However, suppose that we would like to change this program into a game that will *automatically* keep *repeatedly* rolling the die *as long as* the roll is not a 6 (i.e., is a 1, 2, 3, 4, or 5). Once the roll is a 6, the repetition stops, and the game is over.

We can visualize the looping action we need with the following flowchart:

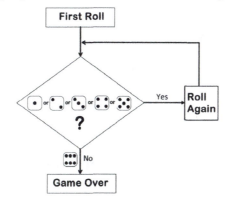

A while statement will enable us to produce this same game action in our program. Let's start by adding to our program the code shown below, which provides the overall structure of the while statement:

```
int roll;
roll = int( random(1, 7) );
println("Roll: " + roll);

while (  )
{

}
```

Next, inside the pair of parentheses, we need to specify the condition that will allow the repetition of the while loop:

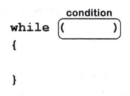

When trying to determine the loop condition that is needed, it can be helpful to paraphrase the word **while** with **"as long as."** In other words, the while loop will continue to perform the statements it contains *as long as* this loop's condition is *true*. For our game example, we want to continue simulating another roll of the die *while* (*as long as*) the current roll is a 1, 2, 3, 4, or 5—in other words, as long as the roll is *not a 6*. We can write this condition for continuing as

```
roll != 6
```

When we update our while statement to use this condition, we now have the following:

```
int roll;
roll = int( random(1, 7) );
println("Roll: " + roll);

while (roll != 6)
{

}
```

Next, between the pair of curly braces, we must specify what is known as the **body** of the while loop.

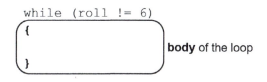

The body of the while loop is where we place any statement(s) that we wish to continue to perform repeatedly *as long as* the loop's condition is *true*. For the game we are developing, these statements are the two that simulate the rolling of the die:

```
int roll;
roll = int( random(1, 7) );
println("Roll: " + roll);

while (roll != 6)
{

}
```

So we will simply copy these two statements and paste copies of them inside the body of the while statement:

```
int roll;
roll = int( random(1, 7) );
println("Roll: " + roll);

while (roll != 6)
{
  roll = int( random(1, 7) );
  println("Roll: " + roll);
}
```

Thus, in our program, we roll the die *once* at the start. Afterward, our while statement determines whether we should *continue* rolling the die. This approach—"do it *once before* we do it over and over again"—is a common way of setting up the use of a while statement.* To help us distinguish the "do it once" part of our program from the "do it over and over again" part of our program, let's slightly modify the console output of each:

```
int roll;
roll = int( random(1, 7) );
println("First roll: " + roll);

while (roll != 6)
{
  roll = int( random(1, 7) );
  println("Roll again: " + roll);
}
```

---

* Some programmers refer to this as "priming the pump."

Let's also add a call to the `println()` function after the `while` statement to announce that the game is over:

```
int roll;
roll = int( random(1, 7) );
println("First roll: " + roll);

while (roll != 6)
{
 roll = int( random(1, 7) );
 println("Roll again: " + roll);
}

println("Game over.");
```

When we run this modified version of our program several times, we see that the *first* roll of the die *always* takes place. After this first roll, the number of times that the simulated rolling of the die is *repeated* by the `while` statement *depends* on how long it takes for a 6 to be rolled. For example, we might see the following console output:

```
First roll: 2
Roll again: 3
Roll again: 5
Roll again: 4
Roll again: 2
Roll again: 6
Game Over!
```

In this particular sample run of our program, after the first roll, the `while` loop *continued* to be performed *five* times before a 6 was rolled, which stopped the looping. The message announcing the end of the game was then displayed to the console.

Why does the roll of a 6 stop the loop? Because a `while` loop continues to perform the statements in its body only *as long as* the loop condition is *true*. The loop condition for this particular `while` statement is

```
roll != 6
```

This condition remains *true* only as long as the roll is *not* a 6. Once a 6 is rolled, this condition becomes *false*, and the looping action of the `while` statement *ends*.

It is important to understand that, *after each time* that the statements in the body of the loop are performed, the loop's condition is evaluated *again* to see if it is still true.

```
while (roll != 6)    still true?
{
    roll = int( random(1, 7) );
    println("Roll again: " + roll);
}
```

If the condition is still true, then all the statements in the body of the loop are performed *again*. Then the loop condition is evaluated to see if it is still true. This cycle of evaluating the loop's test *condition* and performing the statements in the *body* of the loop continues, as long as the condition remains *true*. However, as soon as this condition becomes *false*, the statements in the body of the loop are *not* performed, and the while statement is complete. Processing then moves on to the statement (if any) that immediately follows the while statement. Thus, when we run our current program, the statements in the body of the while loop continue to be performed repeatedly, as long as the condition is true—that is, as long as the roll is not a 6. However, as soon as the value of roll does become a 6, the while statement is complete, and Processing moves on to the next statement—i.e., the one that displays "Game Over!" in the console.

```
First roll: 2
Roll again: 3
Roll again: 4
Roll again: 2
Roll again: 6    ← Roll was a 6, so the while loop ended
Game Over!
```

We cannot predict in advance how many times this particular while loop will be performed. For this reason, this while loop is an example of what is known as an **indefinite** loop. For example, the next time we run our program, we might see the following console output:

```
First roll: 3
Roll again: 1
Roll again: 5
Roll again: 6
Game Over!
```

In this sample run of our program, the while loop was performed only *three* times before a 6 was rolled, stopping the loop.

Sometimes, when we run our program, the *first* roll is a 6. When this happens, we see only the following console output:

```
First roll: 6
Game Over!
```

Because the first roll was a 6, the very *first time* that the condition of the `while` statement was evaluated, it was false. As a result, the statements inside the body of the `while` loop were *not* performed, not even *once*. Rather, Processing skipped over them and moved on to the statement after the `while`.

```
int roll;

roll = int( random(1, 7) );
println("First roll: " + roll);
while (roll != 6 )  ←false!
{
    roll = int( random(1, 7) );
    println("Roll again: " + roll);
}

println("Game Over!");
```

If the first roll is a 6 then the condition is already false.

Thus, the statements in the body of the loop are not performed even once.

Processing then moves on to the first statement after the **while** statement.

The `while` statement is sometimes called a **pretest** or **test-at-the-top** loop because the *condition* is evaluated *before* any of the statements in the body of the loop are executed. Thus, the statements in the *body* of a `while` loop will be performed *zero or more* times.

Save the `RollWhile` program. We will return to this program later in this chapter.

Avoiding Infinite Loops

It is very important that the loop *condition* we choose is one that will eventually become *false*. This ensures that the looping action of the `while` statement will eventually "terminate," that is, stop. Otherwise, if the loop condition never becomes false, then we have created what is known as an **infinite loop**. This is a programming error that, at the very least, will require us to stop the program manually. We might also need to close the Processing program.* It might even be necessary to shut down and restart the computer.

Here's a strategy for avoiding an infinite loop. The loop condition of a `while` statement typically includes a *variable*. For example, the condition of our current `while` statement contains the `roll` variable:

```
while (roll != 6)
```

To prevent an infinite loop, there must be a statement *inside the body* of the loop that *changes* the value of this variable:

---

* In Windows, closing Processing when it is stuck in an infinite loop might require using the Task Manager. In MacOS, this might require using the Force Quit option.

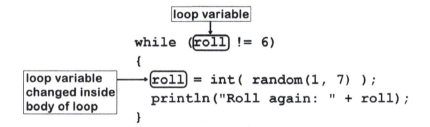

However, having the body of the loop change the value of the loop variable is not enough to prevent an infinite loop. Rather, the value of this loop variable must eventually be changed to a value that causes the loop *condition* to become *false*. For example, in our current game, we have a statement inside the body of the while statement that is changing the value of the roll variable each time the loop is performed:

```
while (roll != 6)
{
  roll = int( random(1, 7) );
  println("Roll again: " + roll);
}
```

We know from our previous experiences with the random() function that this statement will eventually assign the value 6 to the roll variable. When this happens, the loop condition becomes false, and the repeated performance of the statements in the body of the while loop stops.

It is common for beginning programmers to create infinite loops by mistake, so don't feel bad if you accidentally create one. However, it is a good habit to *save* your work *before* running a program that has an untested while loop!

Watch Out for the Extra Semicolon!

As in the case of an if statement, notice that there is no semicolon at the end of the first line of a while statement. In fact, it is even *more* important to avoid this error in a while statement because it might cause an *infinite loop*:

```
while (roll != 6) ;  ←likely to cause an infinite loop!
{
  roll = int( random(1, 7) );
  println("Roll again: " + roll);
}
```

Why might this error cause an infinite loop? Because adding this semicolon at the end of the line essentially creates the equivalent of the following while statement:

```
while (roll != 6)
{
}
```

In other words, we have created a `while` statement that never does anything. Most importantly, the value of our loop variable, `roll`, is never modified inside the body of the loop. Thus, unless the *first* roll is a 6, this loop condition cannot ever become false, and an *infinite loop* will result. Remember *not* to put a semicolon at the end of the first line of a `while` statement!

Using the `while` Statement: A Graphical Example

Let's make a kind of target game using a `while` statement. Select **File > New** to start a new program. Save this program as **BallTarget**.

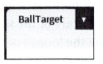

We'll start by creating a canvas and drawing a large square that covers most of the canvas:

```
size(200, 200);
rect(0, 0, 180, 180);
```

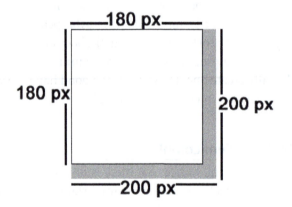

Next, we'll declare two variables to hold the pixel column and pixel row location that we will select at random for each of the little circles we will draw. We'll initialize these variables to zero.

```
size(200, 200);
rect(0, 0, 180, 180);

int column;
int row;

column = 0;
row = 0;
```

Next, we'll add statements that will pick a pixel column number and a pixel row number at random and assign them to the corresponding variables:

```
size(200, 200);
rect(0, 0, 180, 180);

int column;
int row;

column = 0;
row = 0;

column = int( random(width) );
row = int( random(height) );
```

Let's also add a statement to display the pixel row and pixel column that are chosen:

```
size(200, 200);
rect(0, 0, 180, 180);

int column;
int row;

column = 0;
row = 0;

column = int( random(width) );
row = int( random(height) );
println("Column: " + column + " Row: " + row);
```

We will also add two statements to set the fill color to black and to draw a small circle at the selected pixel location:

```
size(200, 200);
rect(0, 0, 180, 180);

int column;
int row;

column = 0;
row = 0;

column = int( random(width) );
row = int( random(height) );
println("Column: " + column + " Row: " + row);
```

```
fill(0, 0, 0);
ellipse(column, row, 10, 10);
```

When we run this program, a circle is drawn at a random location on the canvas,

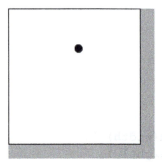

and its pixel column and pixel row are displayed in the console:

```
Column: 96 Row: 53
```

Now, let's use a while loop so that the circles are drawn *repeatedly* on the canvas until a circle's pixel location is outside the square target. For this, we need to wrap a while statement around the four statements that pick the random pixel location and draw the circle:

```
size(200, 200);
rect(0, 0, 180, 180);

int column;
int row;

column = 0;
row = 0;

while ( )
{
  column = int( random(width) );
  row = int( random(height) );
  println("Column: " + column + " Row: " + row);

  fill(0, 0, 0);
  ellipse(column, row, 10, 10);
}
```

What shall we use as the loop *condition* of our while statement? Because the side length of our square is 180 and the radius of the ball is 5, and 180 + 185 equals 185, let's consider the ball to be touching the target as long as the ball's pixel *column* location is less than 185,

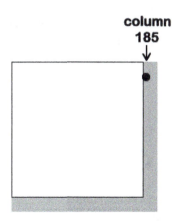

and the ball's pixel *row* location is also less than 185:

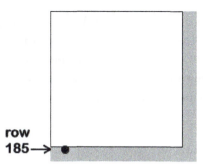

Thus, our loop condition can be

```
(column < 185) && (row < 185)
```

Once we add this loop condition our program is now

```
size(200, 200);
rect(0, 0, 180, 180);

int column;
int row;

column = 0;
row = 0;

while ( (column < 185) && (row < 185) )
{
  column = int( random(width) );
  row = int( random(height) );
  println("Column: " + column + " Row: " + row);
  fill(0, 0, 0);
  ellipse(column, row, 10, 10);
}
```

When we run this program, balls are repeatedly thrown as long as they are at least partially touching the target. However, once one ball completely misses the target, the game is over:

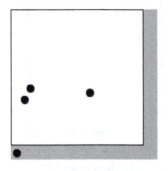

Our while loop's condition contains an AND operation (&&), which specifies that *both* the pixel column and the pixel row must be less than 185 in order for the loop to repeat. Thus, the loop will stop repeating when the pixel column or the pixel row is greater than or equal to 185. In the case of the current sample run of our program, the console output shows that it was the pixel *row* of the last circle that was not less than 185, and this caused the loop to stop:

```
Column: 108 Row: 111
Column: 27 Row: 105
Column: 19 Row: 120
Column: 8 Row: 191
```

It can also be the case that it is the pixel *column* that is not less than 185. This also causes the loop to stop, as in the following sample run:

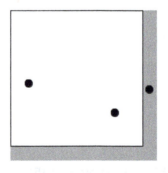

```
Column: 25 Row: 97
Column: 142 Row: 136
Column: 189 Row: 105
```

It can also be the case that *both* the pixel column and the pixel row of the last circle drawn are greater than or equal to 185, as in this sample run of our program:

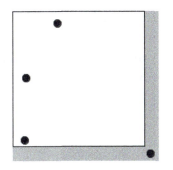

```
Column: 61 Row: 16
Column: 18 Row: 89
Column: 16 Row: 172
Column: 188 Row: 190
```

In this case, *both* conditions joined by the logical AND (&&) operation in our condition are false.

$$\begin{array}{ccc} \text{false} & & \text{false} \\ \texttt{while ((column < 185)} & \texttt{\&\&} & \texttt{(row < 185))} \end{array}$$

However, remember that when two conditions are joined by a logical AND (&&) operation, *both* of the conditions must evaluate to true in order for the result of the && operation to be true. Thus, as soon as the *first* condition turns out to be false, there is no longer any way for the result of the && operation to be true. In fact, Processing does not bother to evaluate the second condition.

This is another example of **short-circuit evaluation** of multiple conditions by Processing.

Let's add a variable to keep track of the number of balls that are thrown until one misses the target.

```
int score;
```

We'll initialize this variable to zero as well:

```
score = 0;
```

We'll also add a call to the `println()` function after the `while` statement to display the final score showing the number of balls that were thrown.

```
println("Score: " + score);
```

How do we increase our `score` variable each time a ball is thrown? For this, we add the following statement inside the body of our `while` statement:

```
score = score + 1;
```

Our program is now

```
size(200, 200);
rect(0, 0, 180, 180);

int column;
int row;
int score;

column = 0;
row = 0;
score = 0;

while ( (column < 185) && (row < 185) )
{
  column = int( random(height) );
  row = int( random(width) );
  println("Column: " + column + " Row: " + row);

  fill(0, 0, 0);
  ellipse(row, column, 10, 10);

  score = score + 1;
}

println("Score: " + score);
```

The assignment statement that we added inside the body of the `while` statement is somewhat unusual in that it has the *same* variable name on *both* sides of the assignment operator (=).

$$\boxed{\text{score}} = \boxed{\text{score}} + 1;$$

Thus, let's take a closer look at this assignment statement. Remember, whenever we have an *expression* on the right-hand side of the assignment operator (=), this expression on the right-hand side is always evaluated *first*.

$$\text{score} = \boxed{\text{score} + 1};$$

For this reason, you may prefer to put a pair of parentheses around the expression on the right-hand side of the assignment operator if you find this makes this assignment statement easier to understand:

```
score = (score + 1);
```

In this particular case, the evaluation of the expression on the right-hand side starts with a retrieval of the copy of the value currently stored in the score variable. Initially, the value stored in score is zero, so we can visualize this as

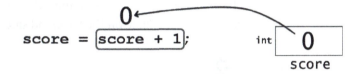

This retrieved value can then be used in the addition operation. We can visualize this operation as

$$0 + 1$$
$$\text{score} = \boxed{\text{score} + 1};$$

After this addition operation is performed, the result is 1. We can visualize the result of this addition as

$$1$$
$$\text{score} = \boxed{\text{score} + 1};$$

As always, when the evaluation of the right-hand side of an assignment statement is complete, the result is stored in the variable on the left-hand side of the assignment operator. Thus, in this particular case, the new value that is now stored in the score variable is 1. We can visualize this assignment operation as

$$1$$
$$\text{score} = \boxed{\text{score} + 1}; \qquad \text{int} \boxed{1}$$
$$\text{score}$$

Thus, the value in the score variable has been increased by 1. This action of increasing the value of the score variable by 1 will be repeated each time the statements in the body of the while statement are performed. For example, if the body of the while statement is performed a second time, we can visualize the steps in the evaluation of the expression on the right-hand side of the assignment statement as

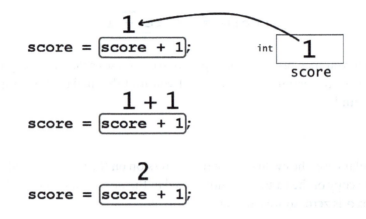

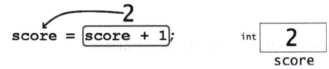

Each time, when the evaluation of the expression on the right-hand side of the assignment operator is complete, the result is stored in the variable on the left-hand side. We can visualize this assignment operation as

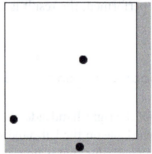

Thus, each time the `while` loop repeats, the value of `score` is increased by 1. As a result, we are able to *count* the number of balls that are thrown.

As before, when we run this program, balls are thrown until one completely misses the target:

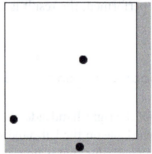

The pixel column and pixel row for each ball are still shown in the console. In addition, the score showing the total number of balls thrown is now also displayed in the console:

```
Column: 107 Row: 76
Column: 12 Row: 155
Column: 102 Row: 192
Score: 3
```

Save and close this **BallTarget** program.*

---

* We could further enhance this program by using `int` variables in place of some of the specific integer values we used: the side length of the square (180), the diameter of the ball (10), and the maximum column and row location (185).

Using the while Statement: Gift Card Simulation

Let's write a program to simulate the use of a $10 gift card at an ice cream shop that sells single, double, and triple cones at the price of $1 per scoop.

Select **File > New** to start a new program. Save this program as **GiftCard**.

We'll need a variable to keep track of the balance on the card:

```
int balance;
```

We'll also need a variable to keep track of the amount of the current expense:

```
int expense;
```

We need to initialize these two variables. We'll start with an opening balance of $10 on the gift card. The initial expense amount will be $0.

```
balance = 10;
expense = 0;
```

Next, let's add a call to the println() function to display the opening balance:

```
println("Opening balance: $" + balance);
```

Thus, our program is currently

```
int balance;
int expense;

balance = 10;
expense = 0;

println("Opening balance: $" + balance);
```

When we run this program, it displays the opening balance in the console:

```
Opening balance: $10
```

Now, let's add statements to simulate the buying of an ice cream cone. First, we'll add a statement that will pick at random an expense (in whole dollars) from $1 to $3 to simulate the choice of a size of ice cream cone (single scoop, double scoop, or triple scoop) and will store this expense in the expense variable:

```
int balance;
int expense;

balance = 10;
expense = 0;

println("Opening balance: $" + balance);

expense = int( random(1, 4) );
```

For example, if the randomly chosen expense is 2 for a double cone, we can visualize the result of this statement as

Next, we'll add a statement to update the gift card balance by subtracting the current expense:

```
int balance;
int expense;

balance = 10;
expense = 0;

println("Opening balance: $" + balance);

expense = int( random(1, 4) );
balance = balance - expense;
```

Once again, the assignment statement we've added is a little unusual in that the same variable name appears on both sides of the assignment operator (=):

$$\boxed{\text{balance}} = \boxed{\text{balance}} - \text{ expense};$$

Let's take a closer look at the effect of this statement. As always, any *expression* on the right-hand side of the assignment operator (=) is always evaluated *first*:

$$\texttt{balance = \boxed{balance - expense};}$$

Again, you may prefer to put a pair of parentheses around the expression on the right-hand side of the assignment operator if you find this makes this assignment statement easier to understand:

```
balance = (balance - expense);
```

When this expression is evaluated, a copy of the value currently stored in the `balance` variable is retrieved. We can visualize this as

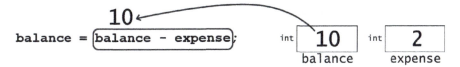

A copy of the value currently stored in the `expense` variable is also retrieved. We can visualize this as

These retrieved values can now be used in the subtraction operation. We can visualize this operation as

The subtraction operation is then performed, producing a result. We can visualize the result of this operation as

Now that the evaluation of the expression on the right-hand side of the assignment statement is complete, the result can be stored in the variable on the left-hand side. Thus, in this particular case, the result becomes the new value of the `balance` variable. We can visualize this assignment operation as

And we can visualize the result of this assignment operation as

Next, let's add a statement that simply outputs both the expense and the new balance to the console:

```
int balance;
int expense;

balance = 10;
expense = 0;

println("Opening balance: $" + balance);

expense = int( random(1, 4) );
balance = balance - expense;
println("expense: $" + expense + " amount remaining: $" + balance);
```

When we run this program, we see that the output to the console displays both the opening gift card balance and the resulting balance after buying one ice cream cone:

```
Opening balance: $10
expense: $2 amount remaining: $8
```

For our simulation, we want to *continue* to perform the last three statements of our program, *as long as* the balance is above $0. Thus, we will put these statements in the body of a while statement:

```
int balance;
int expense;

balance = 10;
expense = 0;

println("Opening balance: $" + balance);

while (balance > 0)
{
  expense = int( random(1, 4) );
  balance = balance - expense;
  println("expense: $" + expense + " amount remaining: $" + balance);
}
```

The statements in the body of this while loop will now continue to be performed as long as the balance is greater than $0.

Let's add a statement after the while loop so that when the balance becomes less than or equal to $0, a message will be displayed in the console announcing that the gift card is used up:

```
int balance;
int expense;

balance = 10;
expense = 0;

println("Opening balance: $" + balance);

while (balance > 0)
{
  expense = int( random(1, 4) );
  balance = balance - expense;
  println("expense: $" + expense + " amount remaining: $" + balance);
}

println("Gift card is used up.");
```

When we run this program, simulations of transactions using the gift card are now displayed in the console until the gift card is used up. Sometimes, this means that the ending balance is $0:

```
Opening balance: $10
expense: $3 amount remaining: $7
expense: $3 amount remaining: $4
expense: $2 amount remaining: $2
expense: $2 amount remaining: $0
Gift card is used up.
```

Other times, the ending balance is less than $0:

```
Opening balance: $10
expense: $1 amount remaining: $9
expense: $2 amount remaining: $7
expense: $3 amount remaining: $4
expense: $2 amount remaining: $2
expense: $3 amount remaining: $-1
Gift card is used up.
```

Whenever the ending balance is negative (less than $0), this means that the customer still owes some money, in addition to the amount that was left on the gift card. To clarify this, we can add a statement that will display an appropriate message in the console:

```
int balance;
int expense;

balance = 10;
expense = 0;

println("Opening balance: $" + balance);

while (balance > 0)
{
  expense = int( random(1, 4) );
  balance = balance - expense;
  println("expense: $" + expense + " amount remaining: $" + balance);
}

println("Gift card is used up.");
println("Amount customer still owes: $" + -(balance) );
```

Notice that we simply used a minus sign (–) to change the negative value of the balance variable to a positive value.

```
println("Amount customer still owes: $" + -(balance) );
```

This – is sometimes called the **unary** minus operator, which we are allowed to use in Processing.* Now, when we run this program and the ending balance is below $0, a message is displayed regarding the amount that the customer still owes:

```
Opening balance: $10
expense: $1 amount remaining: $9
expense: $2 amount remaining: $7
expense: $3 amount remaining: $4
expense: $2 amount remaining: $2
expense: $1 amount remaining: $1
expense: $2 amount remaining: $-1
Gift card is used up.
Amount customer still owes: $1
```

Having console output each time the statements in the body of a while loop are performed helps to clarify the looping action that is taking place. Such console output is known as **tracing** a loop, and it is a good programming habit when working with loops. Such output can greatly simplify the process of determining whether a loop is working correctly and, if not, what changes in the loop are required. Save and close the GiftCard program.

---

* An alternative would be to use Processing's **abs()** function, which returns the absolute value of the given argument:
  println("Amount customer still owes: $" + **abs**(balance));.

Using the while Statement: A Counting Loop to Draw Concentric Circles

Let's use a while statement to draw some concentric circles with gradually increasing diameters. Select **File > New** to start a new program. Save this program as **Concentric**.

```
Concentric  ▼
|
```

We'll set our canvas size to be 150 pixels by 150 pixels:

```
size(150, 150);
```

We'll need a variable to keep track of the changing diameter of the circles. Let's set the initial diameter to 20 pixels.

```
int diameter;
diameter = 20;
```

We will also use a variable named step to keep track of the amount by which the diameter of the circle will increase each time. We'll set the initial step size to 10 pixels. (Using a variable also makes it easy to change this step size, if we ever decide to do so.)

```
int step;
step = 10;
```

Let's turn off the filling of the circles we will draw and set the stroke weight to 3 pixels.

```
noFill();
strokeWeight(3);
```

We'll also add statements to display the diameter to the console and to draw a circle with the current diameter in the center of our canvas:

```
println(diameter);
ellipse(width / 2, height / 2, diameter, diameter);
```

Our program is now

```
size(150, 150);

int diameter;
diameter = 20;

int step;
step = 10;
```

```
noFill();
strokeWeight(3);

println(diameter);
ellipse(width / 2, height / 2, diameter, diameter);
```

When we run this program, we see the current value of our diameter variable output to the console:

20

Also, in the center of the canvas, we see a circle drawn with a diameter equal to the starting value of the `diameter` variable, 20 pixels:

Now, in order to draw increasingly larger circles, we'll need to add a statement to increase the size of the diameter by our chosen step size:

```
println(diameter);
ellipse(width / 2, height / 2, diameter, diameter);
diameter = diameter + step;
```

Once again, we are using an assignment statement that has the same variable name on both sides of the assignment operator (=).

$$\boxed{\texttt{diameter}} = \boxed{\texttt{diameter}} + \texttt{step;}$$

Let's take a closer look at this statement. As always, the *expression* on the right-hand side of the assignment operator (=) is evaluated *first*.

$$\texttt{diameter} = \boxed{\texttt{diameter + step}};$$

Again, you may prefer to put a pair of parentheses around the expression on the right-hand side of the assignment operator if you find this makes this kind of assignment statement easier to understand:

```
diameter = (diameter + step);
```

When this expression is evaluated, a copy of the value currently stored in the `diameter` variable is retrieved. We can visualize this as

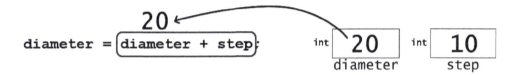

A copy of the value currently stored in the step variable is also retrieved. We can visualize this as

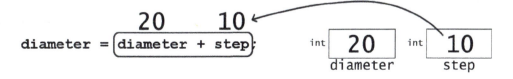

The retrieved values can now be used in addition operation. We can visualize this operation as

$$\overset{\text{20 + 10}}{\text{diameter} = \boxed{\text{diameter + step}}}$$

The addition operation is performed and produces a result. We can visualize the result of this operation as

$$\overset{\text{30}}{\text{diameter} = \boxed{\text{diameter + step}}}$$

As always, once the evaluation of the expression on the right-hand side of an assignment statement is complete, the result is stored in the variable on the left-hand side. Thus, in this particular case, the result becomes the new value of the diameter variable. We can visualize this assignment operation as

$$\text{diameter} = \boxed{\text{diameter + step}}$$

Thus, the diameter is increased by an amount equal to the step size we chose: 10. We can visualize the result of this assignment operation as

Now, in order to *continue* to display the current diameter, draw a circle with the current diameter, and then increase the diameter, we need to wrap these three statements in a while loop:

```
while (  )
{
  println(diameter);
  ellipse(width / 2, height / 2, diameter, diameter);
  diameter = diameter + step;
}
```

What should the *condition* of this while statement be? Remember, it can be helpful to paraphrase the keyword while to "as long as." Here, we wish to keep performing the actions in the body of this loop *as long as* (while) the diameter of our circle is smaller than the canvas. Thus, our condition can be

```
while (diameter < width)
```

With this change, our program is now

```
size(150, 150);

int diameter;
diameter = 20;

int step;
step = 10;

noFill();
strokeWeight(3);

while (diameter < width)
{
  println(diameter);
  ellipse(width / 2, height / 2, diameter, diameter);
  diameter = diameter + step;
}
```

Remember also that, also that in order to avoid an infinite loop, the value of the loop variable used in the condition of our while statement must be *changed* inside the body of the loop in such a way that the loop condition will eventually become false, stopping the loop. In this particular case, the value of the loop variable, diameter, is changing inside the body of the loop: specifically, the value of diameter is increasing by our step size, 10 pixels, each time the body of the loop is performed.

```
while (diameter < width)
{
  println(diameter);
  ellipse(width / 2, height / 2, diameter, diameter);
  diameter = diameter + step;
}
```

Thus, the value of diameter will eventually be greater than the width of the canvas, 150 pixels. At this point, the loop condition

```
while (diameter < width)
```

will become false, and this while loop will end.

Indeed, when we run this program, we see in the console that the value of diameter continues to increase, each time by 10 pixels, only as long as the value of diameter remains less than the width of the canvas, 150 pixels. Once the value of diameter is greater than or equal to 150, the loop condition becomes false, and the looping of the while statement stops.

```
20
30
40
50
60
70
80
90
100
110
120
130
140
```

We also see in the canvas the visual effect of increasing the value of the diameter variable:*

Using a loop to count through a series of numbers like this is an example of what is known as a **counting loop**. (This type of loop is also known as a *definite* loop, because unlike an *indefinite* while or do-while loop, a counting loop will always repeat a specific number of times.) We will learn more about counting loops in Chapter 6.

Save the Concentric program. We will continue working with this program in the next section.

---

* The optical illusion produced by these concentric circles is known as the "moiré effect."

Commenting/Uncommenting

Console output each time the body of a loop is performed is indeed helpful for tracing the repeating actions of a loop. However, if this console output is undesirable once the program is working, these statements can easily be changed to comments simply by putting two forward slashes at the beginning of each line. (Remember: Processing ignores anything from these slashes to the end of the line.) For example, the following change to our program temporarily turns off the displaying of the diameter to the console when the program is run:

```
while (diameter < width)
{
  // println(diameter);
  ellipse(width / 2, height / 2, diameter, diameter);
  diameter = diameter + step;
}
```

This is sometimes known as **commenting out** a line of code, and it is often a much better alternative to *deleting* a line of code. Later, if we wish to reactivate this line of code to trace the loop again, we can simply "uncomment" the line of code by **removing** the two forward slashes at the beginning of the line:

```
while (diameter < width)
{
  println(diameter);
  ellipse(width / 2, height / 2, diameter, diameter);
  diameter = diameter + step;
}
```

In fact, Processing makes it easy to turn on or off the commenting for one or more lines of code simply by selecting the line(s) of code and then choosing **Comment/Uncomment** from the **Edit** menu or by using the keyboard shortcut:

```
Ctrl-/
```

Save and close the Concentric program.

Using the `while` Statement: An Interactive Guessing Game

Let's create a guessing game that lets the player guess a number from 1 to 10.

Select **File > New** to create a new program. Save this program as **Guess**.

As always, in order to do input, we'll need to insert the following import statement that enables us to work with input and message dialogs:

```
import static javax.swing.JOptionPane.*;
```

Remember that the input received from an input dialog is always a String value. Thus, we'll create a String variable to hold the input we receive when the player guesses.

```
import static javax.swing.JOptionPane.*;

String input;
```

We'll also need an int variable to hold the result we obtain when we convert the player's guess from a String value to an int value. For now, we'll initialize this variable to zero:

```
import static javax.swing.JOptionPane.*;

String input;

int guess;
guess = 0;
```

Next, let's declare a variable to hold the secret number and initialize it to a random integer from 1 to 10:

```
import static javax.swing.JOptionPane.*;

String input;

int guess;
guess = 0;

int secret;
secret = int( random(1, 11) );
```

In our input dialog, we'll start by prompting the player to guess a number from 1 to 10. We'll also use a String variable to hold this prompt:

```
import static javax.swing.JOptionPane.*;

String input;

int guess;
guess = 0;

int secret;
secret = int( random(1, 11) );
```

```
String prompt;
prompt = "Guess a number from 1 to 10:";
```

Next, we can add a statement to generate an input dialog box that prompts the player to guess and store whatever `String` value is entered by the player to our `input` variable:

```
import static javax.swing.JOptionPane.*;

String input;

int guess;
guess = 0;

int secret;
secret = int( random(1, 11) );

String prompt;
prompt = "Guess a number from 1 to 10:";

input = showInputDialog(prompt);
```

This dialog box will look like the following:

Remember, the input that we get from an input dialog box is always a `String` value, even if the user enters what looks like a numeric value. Thus, we need to convert this `String` value to an `int` value using the `int()` function and store the result in the `guess` variable:

```
guess = int(input);
```

Our program now is

```
import static javax.swing.JOptionPane.*;

String input;

int guess;
guess = 0;

int secret;
secret = int( random(1, 11) );
```

```
String prompt;
prompt = "Guess a number from 1 to 10:";

input = showInputDialog(prompt);
guess = int(input);
```

When we run the program, we see an input dialog generated. Once the user enters a guess and presses the OK button,

the last statement in our program,

```
guess = int(input);
```

converts the `String` value entered by the user to an `int` value and stores it in the guess variable.

We want to continue to prompt the player and read in a guess as long as the guess is incorrect. Thus, we'll start by surrounding these two statements with a `while` statement:

```
import static javax.swing.JOptionPane.*;

String input;

int guess;
guess = 0;

int secret;
secret = int( random(1, 11) );

String prompt;
prompt = "Guess a number from 1 to 10:";

while (guess != secret)
{
  input = showInputDialog(prompt);
  guess = int(input);
}
```

Now, if we run our program, the dialog boxes continue to appear until the secret number happens to be guessed.

However, this game is currently too hard! Instead, whenever the player guesses incorrectly, we should change our prompt to one suggesting that the player should guess either a *higher* number or a *lower* number. To handle a guess that is too *low*, we can write

```
if (guess < secret)
{
  prompt = "Guess a HIGHER number (1 to 10):";
}
```

We can also add an `else if` clause to handle a guess that is too *high*:

```
if (guess < secret)
{
  prompt = "Guess a HIGHER number (1 to 10):";
}

else if (guess > secret)
{
  prompt = "Guess a LOWER number (1 to 10):";
}
```

We should also add an `else` clause to handle the only remaining possibility—that this latest guess is *correct*:

```
if (guess < secret)
{
  prompt = "Guess a HIGHER number (1 to 10):";
}
else if (guess > secret)
{
  prompt = "Guess a LOWER number (1 to 10):";
}
else
{
  prompt = "Correct! The secret number is: " + secret;
}
```

Thus, our `while` loop is now

```
while (guess != secret)
{
  input = showInputDialog(prompt);
  guess = int(input);
  if (guess < secret)
  {
```

```
    prompt = "Guess a HIGHER number (1 to 10):";
  }
  else if (guess > secret)
  {
    prompt = "Guess a LOWER number (1 to 10):";
  }
  else
  {
    prompt = "Correct! The secret number is: " + secret;
  }
}
```

Thus, we have *nested* an `if` statement inside a `while` statement. When we nest statements like this, it is especially important to keep good alignment of our curly braces. Even so, as the number of enclosed statements grows, we can also add a comment to help us keep track of the purpose of a closing curly brace.

```
} // while
```

This `while` loop will conclude as soon as the player guesses correctly. When this happens, we should display a message dialog to let the player know that this last guess was correct and that the game is now over:

```
showMessageDialog(frame, prompt);
```

Our program is now

```
import static javax.swing.JOptionPane.*;

String input;

int guess;
guess = 0;

int secret;
secret = int( random(1, 11) );

String prompt;
prompt = "Guess a number from 1 to 10:";

while (guess != secret)
{
  input = showInputDialog(prompt);
  guess = int(input);
```

```
if (guess < secret)
{
  prompt = "Guess a HIGHER number (1 to 10):";
}
else if (guess > secret)
{
  prompt = "Guess a LOWER number (1 to 10):";
}
else
{
  prompt = "Correct! The secret number is: " + secret;
}
} // while

showMessageDialog(frame, prompt);
```

When we run this modified version of our program, the player is prompted to make a first guess as before:

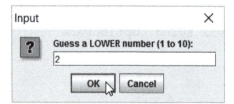

However, if the guess is too high, the player is now prompted to guess a lower number:

Likewise, if the guess is too low, the player is prompted to guess a higher number:

This prompting to the player and reading in a guess continues until the player guesses the correct number, at which point the loop ends, and the user is notified that the guess was correct:

Save and close the Guess program.

The Logical NOT (!) Operator

Processing uses the exclamation point (!) as the logical **NOT** operator. This operator can be used to create the logical *opposite* (also known as the *negation*) of a boolean value, variable, or expression. For example, suppose the value stored in the roll variable is 5. If so, then the boolean expression

```
(roll > 3)
```

evaluates to the boolean value true. The logical NOT operation (!) always produces the logical opposite (negation) of any boolean expression that follows it. Thus, if the value of roll is 5, then the expression

```
!(roll > 3)
```

evaluates to false.

Using Logical NOT (!) With a while Loop's Stop Condition

Processing's while statement asks us to think of a repeating loop in terms of "as long as" rather than in terms of "until." In other words, we need to think in terms of the condition when we want the looping to *continue*.

However, sometimes it easier to think of a particular loop in terms of repeating a certain action *until* a certain condition is true. In such cases, rather than thinking of when we would like the loop to continue, we are thinking instead of the condition when we would like the loop to *stop*. The logical NOT (!) operation can be used to create such while loops.

Let's return to our RollWhile program. Select **File > Save As** and resave this program as **RollWhileNot**.

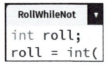

```
RollWhileNot                    ▼
int roll;
roll = int(
```

In the current version of our program, we have defined the `while` loop in terms of the condition under which we would like the performance of the statements in the loop's body to *continue*: specifically, we would like the repetition to continue *as long as* (while) the roll is not equal to 6.

```
int roll;
roll = int( random(1, 7) );
println("First roll: " + roll);

while (roll != 6)
{
  roll = int( random(1, 7) );
  println("Roll again: " + roll);
}

println("Game over.");
```

However, we might be more inclined to think of the loop as repeating *until* a 6 is rolled. If so, then we are thinking of the *stop* condition for the loop rather than of a condition for the loop to *continue*. Specifically, if we would like the loop to stop when a 6 is rolled, then the stop condition we have in mind is

```
roll == 6
```

We can easily convert a *stop* condition into the kind of *continue* condition required by a `while` statement simply by putting the logical NOT (`!`) operator in front of the stop condition. For example,

```
!(roll == 6)
```

If we use this expression as the condition in our `while` statement

```
while ( !(roll == 6) )
```

then this means that the statements in our loop's body will continue to be performed as long as it is *not* the case that the roll is equal to 6. This is exactly what we would like! If we use this modified loop condition, our program is now

```
int roll;
roll = int( random(1, 7) );
println("First roll: " + roll);

while ( !(roll == 6) )
{
  roll = int( random(1, 7) );
  println("Roll again: " + roll);
}

println("Game over.");
```

When we run this program, we see that the loop still repeats, and it stops when a 6 is rolled:

```
First roll: 5
Roll again: 4
Roll again: 1
Roll again: 3
Roll again: 6
Game Over!
```

As the above example illustrates, in cases when it is more intuitive to think of when you would like the loop to *stop*, you can easily use the logical NOT operator (!) to create the logical *opposite* (negation) of the condition under which you want the looping to stop and then use this as the condition of your while statement!

Repeating With the do-while Statement

Processing provides support for another loop structure that is similar to the while loop: the **do-while** statement.

The **do-while** statement is sometimes called a **posttest** or **test-at-the-bottom** loop because the condition that determines whether the loop will continue is performed *after* the statements in the body of the loop have *already* been performed. This means that the statements in the body of the do-while loop will always be performed *at least once*. The basic form of the **do-while** statement is

```
do
{
    statement(s) to perform repeatedly
    as long as condition is true
}
while (condition);
```

where

> *condition* is the condition under which the statement(s) will be repeated.

Notice that a *semicolon* is required at the *end* of a do-while statement.

Let's return again to our RollWhile program. Select **File > Save As** and resave this program as **RollDoWhile**.

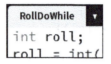

Currently, our program uses a while statement to generate repetition:

```
int roll;
roll = int( random(1, 7) );
println("First Roll: " + roll);

while (roll != 6)
{
  roll = int( random(1, 7) );
  println("Roll again: ", roll);
}

println("Game Over!");
```

To change the while statement in our program to a do-while statement, we can make the following changes:

```
int roll;
roll = int( random(1, 7) );
println("First Roll: " + roll);

do
{
  roll = int( random(1, 7) );
  println("Roll Again: " + roll);
}
while (roll != 6);

println("Game Over!");
```

However, because a do-while statement creates a *posttest* loop (a *test-at-the-bottom loop*), the *body* of the do-while statement is always performed *at least once*. Thus, we no longer need the two statements for the separate *first* roll, as we did before the beginning of the while statement:

```
int roll;
roll = int( random(1, 7) );
println("First Roll: " + roll);
```

Thus, we can change the console output inside the body of the loop to simply "Roll" (instead of "Roll again"):

```
do
{
  roll = int( random(1, 7) );
  println("Roll again: " + roll);
}
while (roll != 6);
```

The do-while version of our program is now

```
int roll;

do
{
  roll = int( random(1, 7) );
  println("Roll: " + roll);
}
while (roll != 6);

println("Game Over!");
```

When we run this program, we see that like the while version, this version still simulates rolling a die *as long as* the roll is not a 6:

```
Roll: 2
Roll: 3
Roll: 3
Roll: 6
Game Over!
```

In this case, using a do-while statement actually simplifies our program somewhat.

In general, the do-while statement is not used as often as the while statement. However, as we see in this example, a do-while statement is sometimes a simpler way of constructing a loop if we know the statements in the loop need to be performed *at least once*.

Summary

In this chapter, we learned about the important principle of *repetition*:

- We learned about creating repetition by looping with the `while` statement.

- We learned about the need to avoid infinite loops.

- We learned that the `while` statement can be used to create an *indefinite* loop that repeats a different number of times each time the program is run.

- We learned that the `while` statement can also be used to create a *definite* loop—also known as a *counting* loop—which repeats a specific number of times each time the program is run.

- We learned about the logical NOT operator (`!`) and how this operator can be used to create `while` loops using *stop* conditions.

- We learned about using the `do-while` statement.

Repetition is one of the most powerful features of any computer programming language.

Exercises

1) Write a program to do the following:

   a) Declare a variable named `age` that can hold an integer.

   b) Assign to `age` a random integer from 1 to 65.

   c) Display "age:" to the console, followed by the value of `age`.

   d) Continue the process of picking and displaying an age as long as the value of `age` is that of an adult (18 or above).

   e) In the console, display the nonadult value of `age` that stopped the loop.

   Sample run

   ```
   age:  28
   age:  20
   age:  43
   age:  56
   age:  13
   13 is not an adult age
   ```

2) Modify the previous program so that the process of picking and displaying an age between 1 and 100 repeats as long as the value of age is instead less than or equal to 90.

Sample output

```
56
63
26
30
28
21
39
96
96 is over 90
```

3) Write a program to do the following:

a) Declare a variable named degrees that can hold an integer.

b) Initialize degrees to 50.

c) Continue to do the following as long as the value of degrees is less than 90:

i) Assign to degrees a random integer from 50 up to 100.

ii) Display the value of degrees to the console followed by "degrees Fahrenheit."

d) Display the last value of degrees followed by "degrees Fahrenheit is hot weather!"

Sample run

```
50 degrees Fahrenheit
50 degrees Fahrenheit
74 degrees Fahrenheit
91 degrees Fahrenheit
91 degrees Fahrenheit is hot weather!
```

4) Write a program to do the following:

a) Declare a variable named amount that can hold an integer.

b) Declare a variable named total that can hold an integer.

c) Initialize total to zero.

d) Continue to perform the following three actions as long as the value of total is less than 20:

    i) Assign to amount a random integer from 1 to 5.

    ii) Display "Amount:" to the console, followed by the value of amount.

    iii) Add amount to total.

e) Display "Total:" to the console, followed by the value of total.

Sample run

```
Amount: 1
Amount: 4
Amount: 1
Amount: 5
Amount: 2
Amount: 2
Amount: 3
Amount: 3
Total: 21
```

5) Write a program to do the following:

a) Set the size of the canvas to 300 pixels by 300 pixels.

b) Declare a variable named `diameter`.

c) Initialize `diameter` to zero.

d) Continue to do the following as long as the value of `diameter` is less than or equal to 100:

   i) Assign to `diameter` a value between 5 and 120.

   ii) Display the value of `diameter` to the console.

   iii) If the value of `diameter` is greater than 100, set the fill color to black; otherwise, set the fill color to white.

   iv) Draw a circle at a random location on the canvas with a diameter equal to the value of `diameter` using the following statement:

```
ellipse(random(width), random(height), diameter,
diameter);
```

Sample run

```
diameter: 40
diameter: 19
diameter: 47
diameter: 110
```

CHAPTER 6

Creating Counting Loops Using the `for` Statement

In computer programming, we sometimes use a loop to *count*. Such a loop is commonly known as a **counting loop**.

Uses of a Counting Loop

As we shall see in this chapter, a counting loop enables us to perform two main kinds of tasks. First, a counting loop enables us to *count through a series of numbers*. For example, suppose we would like to count from 1 to 10 and display these numbers. As we shall see in this chapter, we can use a counting loop for such a task. Second, a counting loop enables us to *repeat a set of statements a specified number of times*. For example, suppose we would like to display "Hello!" to the console 20 times. As we shall see in this chapter, we can use a counting loop for this kind of task as well.

Thus, there are two main reasons for using a counting loop:

1) to *count through a series* of numbers

2) to *repeat* a certain set of statements *a specific number of times*.

Requirements of a Counting Loop

When we create a counting loop, we must meet *three minimum requirements*:

1) *Declare and initialize a* **counting variable**.

2) Define a **loop condition** that uses this counting variable is specified.

3) **Inside the loop,** *change* **the counting variable** in such a way that the loop *condition* eventually becomes *false*, ending the loop.

Let's work through an example of a counting loop that illustrates these three requirements.

Creating a Counting Loop with a `while` Statement

One of the ways that we can create a counting loop is with a `while` statement.

Select **File > New** to create a new program. Save this program as **WhileCount**.

```
WhileCount  ▼
|
```

Enter the following program:

```
int count;
count = 1;

while (count <= 5)
{
  println(count);
  count = count + 1;
}
```

When we run this program, we see the following console output:

```
1
2
3
4
5
```

As we can see, this particular program uses a `while` statement for the first of the two uses of a counting loop that were listed at the beginning of this chapter: *counting through a series of numbers*. This `while` statement also meets the *three requirements* for creating a counting loop. First, this program **declares and initializes a** *counting variable*:

```
int count;
count = 1;

while (count <= 5)
{
  println(count);
  count = count + 1;
}
```

Second, this program has a **loop condition that** *uses the counting variable*:

```
int count;
count = 1;
```

```
while (count <= 5)
{
  println(count);
  count = count + 1;
}
```

Third, **inside the loop, the value of the counting variable is** *changed* in such a way that the loop *condition* eventually becomes *false*, thereby ending the loop:

```
int count;
count = 1;

while (count <= 5)
{
  println(count);
  count = count + 1;
}
```

Here, we see another example of an assignment statement that has the same variable (count) on both sides of the assignment operator (=).

$$\boxed{count} = \boxed{count} + 1;$$

Let's look closer at the action of this statement. The initial value of count is 1. We can visualize this as

int | **1** |
count

As always, an expression on the right-hand side of the assignment operator (=) is evaluated *first*:

$$count = \boxed{count + 1};$$

Again, you may enclose this expression in parentheses if you find that it makes this kind of assignment statement easier to understand:

count = (count + 1);

As the evaluation of this expression begins, a copy of the current value of count is retrieved. We can visualize this as

count = \boxed{count + 1};   int | **1** |
count

The expression on the right-hand side now essentially becomes

$$count = \boxed{1 + 1};$$

The evaluation of the right-hand side can now be completed. We can visualize this as

$$\texttt{count} = \boxed{2};$$

And the result is stored in count. We can visualize this assignment operation as

$$\texttt{count} = 2;$$

We can then visualize the end result of this assignment operation as

int	**2**
	count

Thus, each time the statements in the body of the while loop are performed, the current value of count is output to the console. Then, 1 is added to the current value of the count variable, and the result of this addition is made the *new* value of the count variable. In other words, the value of our counting variable, count, increases by 1 each time the statements in the body of the while statement are performed. Thus, our counting loop is counting upward by 1.

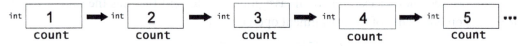

When the body of our while loop is performed the fifth time, the value of count is displayed to the console,

5

and then the value of count increases by 1 to become 6. We can visualize this as

int	**6**
	count

At this point, the loop condition in the while statement is no longer true.

$$\texttt{while (count <= 5)} \leftarrow \textbf{false!}$$

Because the loop condition is no longer true, the repetition of the statements in the body of the while statement *stops*. We can see this even more clearly if we add the following statement to our program:

```
int count;
count = 1;

while (count <= 5)
{
  println(count);
  count = count + 1;
}
```

```
println("Value of count when loop ends: " + count);
```

Now, when we run this program, we see the following output to the console:

```
1
2
3
4
5
Value of count when loop ends: 6
```

Thus, as we have seen, our program meets the three requirements for creating a counting loop:

```
int count;                ← 1) Counting variable declared and initialized.
count = 1;


while (count <= 5)    ← 2) Loop condition that uses counting variable.
{
  println(count);
  count = count + 1; ← 3) Inside the loop, the value of the counting variable
}                            is changed in such a way that the loop condition
                             eventually becomes false, ending the loop.
```

Delete the last statement displaying the ending value of count that we just added to the current program. Then, save the WhileCount program. We will continue working with it in the next section.

Creating a Counting Loop with a `for` Statement

Like most programming languages, Processing provides a special structure for use when a counting loop is needed: the **for** statement.

Recall the three requirements for creating a counting loop:

1) *Declare* and *initialize* a **counting variable**.

2) Define a **loop condition** that uses this counting variable.

3) **Inside the loop,** *change* **the value of the counting variable** in such a way that the loop condition eventually becomes false, ending the loop.

Using a for statement, we can meet all three of these requirement in a *single* statement. The basic form of the **for** statement is

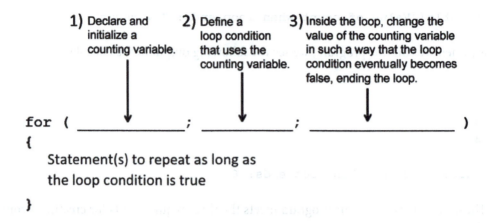

1) Declare and initialize a counting variable.

2) Define a loop condition that uses the counting variable.

3) Inside the loop, change the value of the counting variable in such a way that the loop condition eventually becomes false, ending the loop.

```
for ( _____ ; _____ ; _____ )
{
      Statement(s) to repeat as long as
      the loop condition is true
}
```

Let's try using a `for` statement in place of the `while` statement we used to create our counting loop. Return to the `WhileCount` program. Select **File** > **Save As** and resave this program as **ForCount**.

ForCount ▼

int count

Looking at the current version of our program, remember that we are actually allowed to declare and initialize a variable in a *single* statement. The two statements we are currently using to declare and initialize our variable are

```
int count;
count = 1;
```

These can be replaced with a single statement:

```
int count = 1;
```

Our program is now

```
int count = 1;

while (count <= 5)
{
  println(count);
  count = count + 1;
}
```

This technique of declaring and initializing a variable in a single statement is the form that is typically used in a `for` statement. We can now easily convert our current program to one that uses a `for` statement by changing the code of this program to the following:

```
for (int count = 1; count <= 5; count = count + 1)
{
  println(count);
}
```

When we run this new version of the program, we see the same console output as before:

```
1
2
3
4
5
```

Notice that this for statement does indeed meet all three requirements of a counting loop:

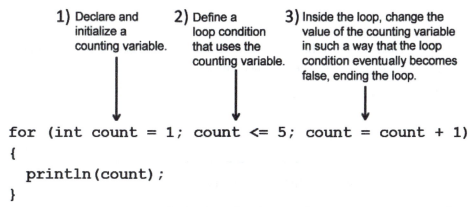

1) Declare and initialize a counting variable.

2) Define a loop condition that uses the counting variable.

3) Inside the loop, change the value of the counting variable in such a way that the loop condition eventually becomes false, ending the loop.

```
for (int count = 1; count <= 5; count = count + 1)
{
   println(count);
}
```

Thus, a for statement can actually be thought of as simply a more *compact* way of writing the same kind of counting loop that we can construct with a while statement.

Save the ForCount program. We will continue working with it in the next sections.

Tracing the Action of the for Statement

The action of the for statement we have written in our ForCount program is the same as that of the counting loop we wrote with a while statement. Let's *trace* this action so that we can understand it more clearly.

Our for statement starts by declaring the variable count and initializing it to 1.

```
for (int count = 1; count <= 5; count = count + 1)
{
   println(count);
}
```

int | 1 |
count

Next, the loop condition is checked to see if it is true. If it is false, then the body of the for statement will not be performed even once. However, in our current for statement, the loop condition is indeed true.

true

```
for (int count = 1; count <= 5; count = count + 1)
{
    println(count);
}
```

int [1]
count

Thus, the body of the for statement is performed for the first time.

```
for (int count = 1; count <= 5; count = count + 1)
{
    println(count); ←——
}
```

int [1]
count

This produces the following console output:

```
1
```

Once the performance of the body of the for statement is complete, the value of the count variable is increased by the specified amount (by 1, in this case).

```
for (int count = 1; count <= 5; count = count + 1)
{
    println(count);
}
```

int [2]
count

Remember, it's in the first line of the for statement,

```
for (int count = 1; count <= 5; count = count + 1)
```

where we specify how the counting variable will change after each pass through the loop. This specification controls the changing of the counting variable in a for statement. Thus, we should *not change* the value of the counting variable by any other means. For example, *no* statement in the *body* of the for statement should change the value of the counting variable, because this is likely to produce unexpected results.

After the loop variable is changed, the loop condition is checked again, to see if it is still true:

still true?

```
for (int count = 1; count <= 5; count = count + 1)
{
    println(count);
}
```

int [2]
count

In our example, the condition is indeed still true, so the body of the for statement is performed a second time:

```
for (int count = 1; count <= 5; count = count + 1)
{
    println(count);  ←——
}
```

int | 2
count

This produces a second line of console output:

```
1
2 ←
```

Afterward, the count variable is increased again by 1.

```
for (int count = 1; count <= 5; count = count + 1)
{
    println(count);
}
```

int | 3
count

This cycle continues until the value of count is 5.

```
for (int count = 1; count <= 5; count = count + 1)
{
    println(count);  ←——
}
```

int | 5
count

This value of count is output to the console:

```
1
2
3
4
5 ←
```

Next, the count variable is incremented again, and its value is now 6.

```
for (int count = 1; count <= 5; count = count + 1)
{
    println(count);
}
```

int | 6
count

At this point, the condition is checked again and is now found to be false.

false!
↓

```
for (int count = 1; count <= 5; count = count + 1)
{
    println(count);
}
```

int | 6
count

Thus, the statements in the body of this `for` loop will not be performed again. The performance of this `for` statement is now complete.

Caution: Only Two Semicolons!

Notice that in the first line of the for statement in our ForCount program, only *two* semicolons are used to separate the three required parts contained within the parentheses:

```
for (int count = 1; count <= 5; count = count + 1)
{
    println(count);
}
```

A semicolon is *not* used after the *third* part, and using one will cause an error:

ERROR
↓

```
for (int count = 1; count <= 5; count = count + 1;)
{
    println(count);
}
```

Also, as in the case of the if and while statements, be careful *not* to put a semicolon at the *end* of the first line of the for statement:

ERROR!
↓

```
for (int count = 1; count <= 5; count = count + 1);
{
    println(count);
}
```

Avoiding Infinite Loops

It is important when using a for statement to avoid inadvertently creating an infinite loop. Given the *initial* value of the counting variable, the specified change in the counting variable that is made each time the loop is repeated must be such that the *condition* of the for statement eventually becomes *false*. Otherwise, an infinite loop will result.

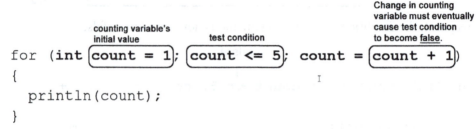

```
for (int count = 1; count <= 5; count = count + 1)
{
    println(count);
}
```

In the for statement in the current version of our ForCount program, the value of the loop variable starts at 1, and it increases by 1 each time the loop is repeated. Thus, we know that the value of the loop variable will eventually become greater than 5. At that point, the loop condition will become false, ending the repetition of the loop.

Incrementing and Decrementing the `for` Loop Counter Variable

Return to the ForCount program. As we have learned, a very important part of any counting loop is the statement that changes the value of the loop counter variable in such a way that the loop condition will eventually become false, which stops the repetition of the loop. This change in the loop counter variable is often by means of an *addition* operation that *increases* its value. This is known as **incrementing** the loop counter variable:

```
for (int count = 1; count <= 5; count = count + 1)
{
  println(count);
}
```

In our current `for` statement, we are incrementing the loop counter variable by 1, but we can also increment by other amounts. For example, let's instead increase the value of our counting variable by 2 each time the `for` loop is performed:

```
for (int count = 1; count <= 5; count = count + 2)
{
  println(count);
}
```

When we run this program, we can see from the console output that the `for` loop now counts from 1 to 5 in increments of 2:

```
1
3
5
```

As another example, let's make the following changes to the initial value of the counting variable, the increment, and the loop condition:

```
for (int count = 0; count <= 50; count = count + 10)
{
  println(count);
}
```

When we run this program, we can see from the console output that the `for` loop now counts from 0 to 50 in increments of 10:

```
0
10
20
30
40
50
```

We can also count *downward* in a counting loop by *decreasing* the value of our loop counter variable. For this, we need to initialize our counting variable to the *maximum* value in our series of numbers. For example,

```
for (int count = 5;          )
```

We also need to change our loop condition so that it checks whether the value of our counting variable is greater than or equal to (or simply greater than) the *minimum* value in our series of numbers, for example,

```
for (int count = 5; count >= 0;      )
```

Importantly, we use a *subtraction* operation in the portion of our `for` statement that modifies the loop counter variable. This is known as **decrementing** the loop counter variable.

```
for (int count = 5; count >= 0; count = count - 1)
```

If we make these changes, then our program is now the following:

```
for (int count = 5; count >= 0; count = count - 1)
{
  println(count);
}
```

When we run this program, we see that the `for` loop now counts from 5 down to 0 by ones:

```
5
4
3
2
1
0
```

Similarly, if we make the following changes to our program,

```
for (int count = 50; count >= 0; count = count - 10)
{
  println(count);
}
```

then when we run this program, we see that the `for` loop counts from 50 down to 0 by tens:

```
50
40
30
20
10
0
```

Save the `ForCount` program. We will continue using it in the next section.

Using Equals in the Loop Condition

It is important to notice that omitting or including the *equals sign* (=) in the loop condition of a counting loop often changes the number of times that loop is performed. Return to the ForCount program. Let's make the following changes to the for statement in this program:

```
for (int count = 1; count < 5; count = count + 1)
{
  println(count);
}
```

When we run this program, we see that it now produces the following console output:

```
1
2
3
4
```

In this case, because we used strictly the less than operator (<) in our loop condition, the loop condition became *false* when the value of the counter variable reached the value in our loop condition, 5. Thus, the body of the for statement was only performed four times, and the value used in our loop condition, 5, was not displayed. However, if we add an equals sign to the loop condition,

```
for (int count = 1; count <= 5; count = count + 1)
{
  println(count);
}
```

then in this case, the loop condition does not become false until *after* the value of the counting variable reaches the value in our loop condition, 5. In this case, the loop condition became false only when the value of the counting variable became 6. Thus, the value in our loop condition was displayed in the console output:

```
1
2
3
4
5
```

As we see here, adding the equals sign (=) to the loop condition of a counting loop often results in an extra repetition of the statements in the body of that loop.

Thus, when using a counting loop to count through the series of numbers, if the value used in the loop condition is one that we want the loop to reach (in this case, 5), then we *include* the equals sign. On the other hand, if the value used in the loop condition is an upper limit that we do *not* want to be used by the statements in the body of the loop, then we *omit* the equals sign.

Save the ForCount program. We will continue working with it in the next section.

Repeating Actions with a Counting Loop

We began this chapter by suggesting that a counting loop enables us to perform two main types of tasks:

1) Counting through a series of numbers

2) Repeating a set of statements a specific number of times

We have already seen how a for statement enables us to create a counting loop that performs the first type of task: *counting through a series of numbers*. Next, let's look at how a for statement can also be used to create a counting loop that performs the second type of task: *repeating a set of statements a specific number of times*.

Return to the ForCount program. Change the for statement in this program to the following:

```
for (int count = 1; count <= 3; count = count + 1)
{
  println(count);
}
```

When we run this modified version of our program, we see that the for statement now counts from 1 through 3, outputting each of the values in this series to the console:

```
1
2
3
```

Here's a reliable strategy for creating a counting loop to perform a certain set of statements a specific number of times:

1) *Initialize* the counting variable to 1.

2) Create a loop condition so that the loop will repeat as long as the counting variable is *less than or equal to* the number of times the loop is to be performed.

3) *Increase* the value of the counting variable by 1 each time the loop is performed.

Our current for statement illustrates this strategy:

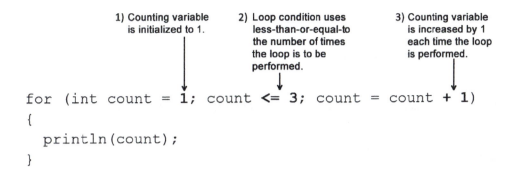

```
for (int count = 1; count <= 3; count = count + 1)
{
    println(count);
}
```

The first line of our `for` statement causes the value of our counting variable, count, to start at 1 and then change to 2, and then to 3, and then stops the loop when the value of count reaches 4. Thus, when we run this program, we see the following output to the console:

```
1
2
3
```

Now, instead of simply displaying the current value of count to the console each time, let's instead output "Hello", along with the value of the counting variable. For this, we simply make the following change to the `for` statement:

```
for (int count = 1; count <= 3; count = count + 1)
{
    println("Hello " + count);
}
```

When we run this program, we now see the following output to the console:

```
Hello 1
Hello 2
Hello 3
```

Thus, if we would like to simply display "Hello" to the console *three* times without displaying the current value of count, we could simply remove the reference to count from our call to the `println()` function.

```
for (int count = 1; count <= 3; count = count + 1)
{
    println("Hello " + count);
}
```

As a result, our program is now

```
for (int count = 1; count <= 3; count = count + 1)
{
  println("Hello ");
}
```

When we run this program, we now see the following output to the console:

```
Hello
Hello
Hello
```

Thus, our counting loop is now being used to repeat a certain statement a specified number of times. Our counting variable, count, is still being used to drive this counting loop. The only difference here is that now we are not making any visible use of the changing values of this counting variable, because we are no longer outputting the value of count to the console each time. If we would like to output "Hello" to the console *five* times, then we can simply change our for statement to

```
for (int count = 1; count <= 5; count = count + 1)
{
  println("Hello ");
}
```

When we run this program, it now generates the following output to the console:

```
Hello
Hello
Hello
Hello
Hello
```

Thus, we have now seen that the for statement works very well for the two main uses of a counting loop:

1) To *count through a sequence of numbers*

2) To have a particular set of *statements repeated a specific number of times*

Save the ForCount program. We will continue working with it in the next section.

Scope of the for Statement's Counting Variable

Let's return to the ForCount program and make the following change to the for statement:

```
for (int count = 1; count <= 5; count = count + 1)
{
  println(count);
}
```

When the variable used for counting is declared in the first line of a `for` statement, as it is in our current program, this counting variable *cannot* be used *outside* that `for` *statement*.

The range of a program where a particular variable may be used is sometimes called the **scope** of that variable. The scope may be thought of as the "life span" of the variable: after we have moved beyond the range of the program where this variable may be used, this variable is no longer available for us to use. (We'll learn more about variable scope later in this book.) For example, in the following program, the `count` variable may only be used within the `for` statement in which it is declared.

```
for (int count = 1; count <= 5; count = count + 1)
{
  println(count);
}
```
the scope where count may be used

Now, let's try adding the following statement:

```
for (int count = 1; count <= 5; count = count + 1)
{
  println(count);
}

println("Value of count when loop ends: " + count);
```

Processing underlines the reference to `count` in this added statement:

```
println("Value of count when loop ends: " + count);
```

Also, the following error message is displayed in the Message Area:

The variable "count" does not exist

This error results because the last statement attempts to use the variable `count` *outside* the variable's scope, outside the range of the program where this variable can be used.

```
for (int count = 1; count <= 5; count = count + 1)
{
  println(count);
}
```
the scope where count may be used

```
println("Value of count when loop ends: " + count);
```
↑
outside the scope where count may be used

Delete the statement that we just added. Save and close the `ForCount` program.

Counting through a Sequence with a `for` Statement: Grayscale Example

As we have learned, the RGB color model uses a trio of numbers to define a particular color: a red amount, a green amount, and a blue amount. Each of these numbers is a value from 0 to 255. The RGB numbers for black are 0, 0, 0. The RGB numbers for white are 255, 255, 255. Thus, for black and white, *all three* RGB numbers are the *same*. As it turns out, any other set of RGB numbers in which all three values are the *same* describes some shade of *gray*. For example, given that 127 is approximately half of 255, the RGB numbers 127, 127, 127 define medium gray, about halfway between black and white on what is known as a *grayscale*.

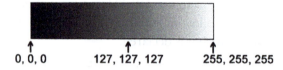

Let's use a `for` statement to draw a similar grayscale. Select **File > New** to start a new program. Save this program as **Grayscale**.

We'll start by setting the size of our canvas:

```
size(256, 50);
```

We have specified a canvas with a width of 256 pixels.

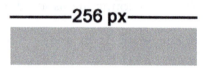

Thus, the pixel column numbers will run from 0 to 255.

Adding the following `for` statement,

```
size(256, 50);

for (int count = 0; count <= 255; count = count + 1)
{
  println(count);
}
```

will display the values from 0 to 255 in the console:

```
0
1
2
3
⋮
252
253
254
255
```

Each time this loop repeats, let's draw a vertical line segment from the top of the canvas (row 0) to the bottom of the canvas (row 49):

We'll draw this line segment at the pixel column position equal to the value of the counting variable. To achieve this, we'll insert the following call to the `line()` function in the body of our `for` statement:

```
size(256, 50);

for (int count = 0; count <= 255; count = count + 1)
{
  println(count);
  line(count, 0, count, 49);
}
```

When the value of `count` starts at zero, the following line segment will be drawn:

And, as the value of `count` progresses toward 255,

another line segment will be added at each pixel column, all the way to the right end of the canvas. Thus, when we run our program, we see the same console output as before:

```
0
1
2
3
  ⋮
252
253
254
255
```

However, the canvas output now looks like a single black rectangle:

This black rectangle results because a black vertical line segment, one pixel wide, has been drawn from the top to the bottom of the canvas in *every* pixel column, from column 0 through column 255.

Recall that the stroke() function sets the stroke color that will be used to draw a point, line segment, or shape edge. Thus, as we count from 0 to 255, we can set all three of the RGB values of the stroke color to the current value of the count variable by inserting the following statement:

```
size(256, 50);

for (int count = 0; count <= 255; count = count + 1)
{
  println(count);
  stroke(count, count, count);
  line(count, 0, count, 49);
}
```

As a result, the stroke color will no longer be black each time a vertical line segment is drawn in a pixel column. Rather, the stroke color starts at black (0, 0, 0) when count is 0 and moves gradually toward white (255, 255, 255) as count progresses to 255. Because the current value of count is used for all *three* RGB numbers, each line segment that is added at each pixel column is the next shade *lighter* on the grayscale than the line segment in the preceding pixel column. Thus, when we run this program, we see the same console output as before,

```
0
1
2
3
```

⋮

252
253
254
255

but we now see a grayscale drawn on the canvas:

This grayscale is comprised of 256 vertical line segments, each of which is the next lighter shade on the grayscale.

Save and close the Grayscale program.

Counting through a Sequence with a for Statement: Revisiting an Example

Let's return to the Concentric program from Chapter 5. Select **File > Save As** and resave this program as **ConcentricFor**.

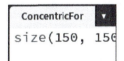

In this program, we created a counting loop using a while statement. When we run this program, we see that this counting loop draws concentric circles on the canvas

and displays each of the diameters of these circles in the console:

20
30
40
50
60
70
80
90
100
110
120

```
130
140
```

Our current program that produces this output consists of the following. Notice that this program includes all three of the characteristics of a counting loop:

```
size(150, 150);

int diameter;
diameter = 20;                          1) declares and initializes a counting variable

int step;
step = 10;

noFill();
strokeWeight(3);

while (diameter < width)                2) has a loop condition that uses the counting variable
{
  println(diameter);
  ellipse(width / 2, height / 2, diameter, diameter);
  diameter = diameter + step;           3) changes the value of the counting variable inside the loop in
}                                          such a way that the loop condition will eventually become false,
                                           ending the loop
```

Because a for statement provides a more compact way of writing the counting loop we created with a while statement, we can easily make the following changes:

```
size(150, 150);

int step;
step = 10;

noFill();
strokeWeight(3);

for (int diameter = 20; diameter < width; diameter = diameter + step)
{
  println(diameter);
  ellipse(width / 2, height / 2, diameter, diameter);
}
```

When we run this modified version of the program, we see that the diameters displayed in the console are the same as before:

```
20
30
40
```

```
50
60
70
80
90
100
110
120
130
140
```

Also, the concentric circles drawn on the canvas are the same as before:

If we would like to randomize the stroke color of each of these circles, we can insert the following statement:

```
stroke( random(256), random(256), random(256) );
```

Each of the nested calls to the random() function returns a random float value from 0.0 up to (but not including) 256.0. As in the case of most of Processing's drawing functions, the three RGB numbers we supply as arguments to the stroke() function are expected to be float values, but any decimal portions of these values are essentially ignored by this function. Thus, there is no need for us to convert these random float values to int values. Our program is now

```
size(150, 150);

int step;
step = 10;

noFill();
strokeWeight(3);

for (int diameter = 20; diameter < width; diameter = diameter + step)
{
  println(diameter);
  stroke( random(256), random(256), random(256) );
  ellipse(width / 2, height / 2, diameter, diameter);
}
```

When we run our program, we see the same console output as before, but our circles now have random stroke colors:

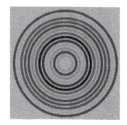

Save and close the ConcentricFor program.

Performing Actions a Specified Number of Times with a for Statement: Coin Flip Simulation Example

We have seen that one of the common uses of a counting loop is to *count through a series of numbers* such as 0–255. However, we have also seen that another common use of a counting loop is to *perform a certain set of actions a specified number of times*. For example, suppose we would like to simulate the flipping of a coin a certain number of times to see how many times "heads" is the result and how many times "tails" is the result Let's write a program to do this.

Select **File > New** to start a new program. Save this program as **CoinToss**.

We will need a variable to keep track of the number of flips of a coin that we would like to simulate. For example, if we intend to simulate the flipping of the coin 10 times, we can write

```
int total;
total = 10;
```

We also need variables to keep track of the number of heads and the number of tails that result. Since we haven't started counting yet, we'll initialize these variables to zero:

```
int total;
total = 10;
```

```
int heads;
heads = 0;

int tails;
tails = 0;
```

We can simulate a flip of a coin by picking a 1 or a 2 at random, with a 1 counting as a "heads" result and a 2 counting as a "tails" result. Thus, we will also need a variable in which to store this random value representing the coin flip:

```
int total;
total = 10;

int heads;
heads = 0;

int tails;
tails = 0;

int flip;
```

Next, let's construct our for statement. In order to count from 1 up to the total number of times we wish to flip, we can write

```
int total;
total = 10;

int heads;
heads = 0;

int tails;
tails = 0;

int flip;

for (int count = 1; count <= total; count = count + 1)
{

}
```

In the body of a for statement, we specify the actions that we wish to repeat. First, to simulate the toss of the coin, we will pick a 1 or a 2 at random and store the result in the flip variable:

```
for (int count = 1; count <= total; count = count + 1)
{
  flip = int( random(1, 3) );
}
```

Next, each time a value is chosen for flip, we need to check if it is 1 or 2. If the resulting value of flip is 1, we will consider this to be "heads." In this case, we will display "heads" to the console and increment our heads variable by 1:

```
for (int count = 1; count <= total; count = count + 1)
{
  flip = int( random(1, 3) );

  if (flip == 1)
  {
    println("heads");
    heads = heads + 1;
  }
}
```

heads

Otherwise, the value assigned to flip is 2, which we consider to be "tails." In this case, we will display "tails" to the console and increment our tails variable by 1:

```
for (int count = 1; count <= total; count = count + 1)
{
  flip = int( random(1, 3) );

  if (flip == 1)
  {
    println("heads");
    heads = heads + 1;
  }

  else // flip is 2
  {
    println("tails");
    tails = tails + 1;
  }
}
```

heads

tails

Because we are nesting an if-else statement inside the for statement, it is important to keep our curly braces aligned and indented appropriately. We can also add a comment identifying the curly brace closing the for statement:

```
for (int count = 1; count <= total; count = count + 1)
{
  flip = int( random(1, 3) );
  if (flip == 1)
  {
    println("heads");
    heads = heads + 1;
  }
```

```
 else // flip is 2
 {
   println("tails");
   tails = tails + 1;
 }
} // for
```

After our `for` statement has completed its repetitions, we need only to display the resulting totals. Adding the following lines of code to our program will acheive this:

```
println(total + " flips");
println(heads + " heads");
println(tails + " tails");
```

Thus, our program is now

```
int total;
total = 10;

int heads;
heads = 0;

int tails;
tails = 0;

int flip;

for (int count = 1; count <= total; count = count + 1)
{
  flip = int( random(1, 3) );

  if (flip == 1)

  {
    println("heads");
    heads = heads + 1;
  }

  else
  {
    println("tails");
    tails = tails + 1;
  }

} // for

println(total + " flips");
println(heads + " heads");
println(tails + " tails");
```

When we run this program, we see from the console output that each time the body of the for statement is performed, "heads" or "tails" is chosen. After this for loop is completed, the total number of flips, the total number of heads, and the total number of tails are also displayed:

```
heads
tails
heads
tails
tails
heads
heads
heads
heads
tails
10 flips
6 heads
4 tails
```

If we wish to change the number of simulated coin flips to 100, we can simply change the initial value assigned to the total variable:

```
total = 100;
```

Now, when we rerun the program, we see the results of 100 simulated coin flips:

```
⋮
tails
heads
heads
100 flips
52 heads
48 tails
```

We can just as easily specify 1000 coin flips,

```
total = 1000;
```

and Processing still delivers the results very quickly:

```
⋮
tails
heads
heads
1000 flips
512 heads
488 tails
```

We see here the power of the computing principle of **repetition**. Humans find repetition to be tedious, but computers perform repetitions very well.

Save and close the `CoinToss` program.

Performing Actions a Specified Number of Times with a `for` Statement: Starry Night Example

We have seen that a common use of a counting loop is to perform a certain action a specified number of times. Using logic similar to that of our coin-flipping program, we can generate a starry sky. Select **File > New** to start a new program. Save this program as **Stars**.

We'll start with a 300-pixel by 200-pixel canvas with all the pixels set to black:

```
size(300, 200);
background(0, 0, 0);
```

Let's also define a variable to specify the number of stars we would like to draw:

```
size(300, 200);
background(0, 0, 0);

int stars;
stars = 10;
```

Next, we'll construct a `for` statement that will repeat once for each star we would like:

```
size(300, 200);
background(0, 0, 0);

int stars;
stars = 10;

for (int count = 1; count <= stars; count = count + 1)
{

}
```

In the body of the `for` loop, we specify the actions we would like to have repeated. We'll start by setting the stroke color to white:

```
size(300, 200);
background(0, 0, 0);
```

```
int stars;
stars = 10;

for (int count = 1; count <= stars; count = count + 1)
{
  stroke(255, 255, 255);
}
```

Next, we'll use the point() function to draw a point that is 1 pixel in size on the canvas. (Processing's default stroke weight is 1 pixel.) Recall that the two arguments that the point() function needs us to supply are simply the pixel column and the pixel row where we would like the point drawn:

```
point(column, row);
```

As in the case of most drawing functions, these arguments of the point() function are expected to be float values, even though the decimal portions of these values are essentially ignored. However, this means that, we can pick the pixel column argument for the point() function at random without having to convert the float result returned by the random() function to an int value:

```
point( random(width),
```

We can similarly pick the pixel row argument for the point() function at random:

```
point( random(width), random(height) );
```

Thus, each time the body of the for statement is repeated, a one-pixel white "star" will be drawn at a randomly chosen pixel location on the canvas. Our program is now

```
size(300, 200);
background(0, 0, 0);

int stars;
stars = 10;

for (int count = 1; count <= stars; count = count + 1)
{
  stroke(255, 255, 255);
  point( random(width), random(height) );
}
```

When we run this program, we see 10 stars drawn at random pixel locations on the canvas:

If we change the initial value of the stars variable to 100,

```
stars = 100;
```

then we now see 100 stars drawn on the canvas:

Simply by changing the value of the stars variable,

```
stars = 1000;
```

we can see as many stars on the canvas as we would like:

If we would like the user to be able to specify the number of stars drawn, we can make the following changes to our program:

```
import static javax.swing.JOptionPane.*;

size(300, 200);
background(0, 0, 0);
```

```
String input;
input = showInputDialog("Enter the number of stars:");

int stars;
stars = int(input);

for (int count = 1; count <= stars; count = count + 1)
{
  stroke(255, 255, 255);
  point( random(width), random(height) );
}
```

Now, when we run our program, the user can enter a number of stars,

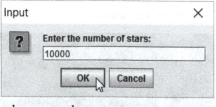

and this many stars will be drawn on the canvas:

Save and close the Stars program.

Counting Through a Sequence with a for Statement: Calculating a Factorial

The **factorial** of a number (assuming the number is a nonnegative integer) is the product of all the positive integers less than or equal to that number. For example, the factorial of 5—written **5!**—is

$$1 \times 2 \times 3 \times 4 \times 5 = 120$$

We can use a for statement to calculate a factorial. Select **File > New** to start a new program. Save this program as **Factorial**.

We will need a variable to store the nonnegative integer for which we are calculating the factorial:

```
int integer;
```

Let's initialize this variable to 5 so that 5! will be calculated:

```
int integer;
integer = 5;
```

We also need a variable to store the factorial that is being calculated. As we shall see, we need this variable to be initialized to 1.

```
int integer;
integer = 5;

int factorial;
factorial = 1;
```

Next, we need a `for` statement that will count from 1 up to the number for which we are calculating the factorial. To test this counting loop, let's start by displaying in the console the changing value of our counting variable:

```
int integer;
integer = 5;

int factorial;
factorial = 1;

for (int count = 1; count <= integer; count = count + 1)
{
    println(count);
}
```

When we run this program, we see the following output to the console:

```
1
2
3
4
5
```

Our `for` loop is counting from 1 to 5, so it's ready to use for calculating the factorial of 5. Each time the body of the `for` statement is performed, we want to take the value currently stored in `factorial` and multiply this value by the current value of `count`. We'll then take this result and make it the new value of the `factorial` variable.

```
int integer;
integer = 5;
```

```
int factorial;
factorial = 1;

for (int count = 1; count <= integer; count = count + 1)
{
   factorial = factorial * count;
}
```

Let's take a closer look at this assignment statement. Remember, an *expression* on the right-hand side of the assignment operator (=) is always evaluated *first*:

$$\text{factorial} = \boxed{\text{factorial * count}};$$

Once again, you may enclose this expression in parentheses if it makes this kind of assignment statement easier for you to understand:

```
factorial = (factorial * count);
```

In this case, the expression on the right-hand side of the assignment statement consists of taking the current value of factorial and multiplying it by the current value of count, producing some result. We can visualize this as

result

$$\text{factorial} = \boxed{\text{factorial * count}};$$

This result is then stored as the new value of the factorial variable. We can visualize this assignment operation as

result

$$\text{factorial} = \boxed{\text{factorial * count}};$$

For example, the initial value of factorial is 1, and the initial value of count is also 1. Thus, the expression on the right-hand side essentially becomes 1 * 1.

The expression 1 * 1 evaluates to 1. Thus, 1 is stored as the new value of factorial.

$$\text{factorial} = \boxed{\text{factorial * count}}; \quad \text{int}\ \boxed{1}\ \text{int}\ \boxed{1}$$
count   factorial

Next, the value of count increases to 2. Thus, the expression on the right-hand side is 1 * 2.

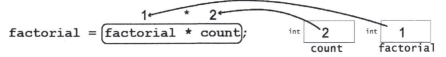

The expression 1 * 2 evaluates to 2. Thus, 2 is stored as the new value of factorial.

factorial = factorial * count;   int [2] count   int [2] factorial

Next, the value of count increases to 3. Thus, the expression on the right-hand side is 2 * 3.

factorial = factorial * count;   int [3] count   int [2] factorial

The expression 2 * 3 evaluates to 6. Thus, 6 is stored as the new value of factorial.

factorial = factorial * count;   int [3] count   int [6] factorial

Next, the value of count increases to 4. Thus, the expression on the right-hand side is 6 * 4.

factorial = factorial * count;   int [4] count   int [6] factorial

The expression 6 * 4 evaluates to 24. Thus, 24 is stored as the new value of factorial.

factorial = factorial * count;   int [4] count   int [24] factorial

Next, the value of count increases to 5. Thus, the expression on the right-hand side is 24 * 5.

factorial = factorial * count;   int [5] count   int [24] factorial

The expression 24 * 5 evaluates to 120. Thus, 120 is stored as the new value of factorial.

Next, the value of count increases to 6. Our loop condition is now false, so the looping stops. The performance of the for statement is complete.

Notice that each time the body of the `for` statement is performed, the resulting value of `factorial` is indeed equal to the factorial of the current value of `count`. Thus, when the loop stopped, the final value of `factorial` was equal to the factorial of 5.

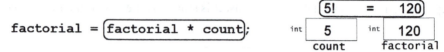

We can display each of these calculated factorials by adding the following call to the `println()` function inside the body of our loop:

```
int integer;
integer = 5;

int factorial;
factorial = 1;

for (int count = 1; count <= integer; count = count + 1)
{
  factorial = factorial * count;
  println(count + "! = " + factorial);
}
```

Also, after the `for` statement, let's add another call to the `println()` function to display the final result:

```
int integer;
integer = 5;

int factorial;
factorial = 1;

for (int count = 1; count <= integer; count = count + 1)
{
  factorial = factorial * count;
  println(count + "! = " + factorial);
}

println("Answer: " + integer + "! is " + factorial);
```

When we run this program, we now see the following output to the console:

```
1! = 1
2! = 2
3! = 6
4! = 24
5! = 120
Answer: 5! is 120
```

This output is correct. And if we change the initial value of integer to 6,

```
integer = 6;
```

then when we run our program, we see the following output:

```
1! = 1
2! = 2
3! = 6
4! = 24
5! = 120
6! = 720
Answer: 6! is 720
```

This output is also correct. Now, let's change our initial value of the integer variable to 1:

```
integer = 1;
```

Now, when we run the program, we see the following output in the console:

```
1! = 1
Answer: 1! is 1
```

This output is also correct. Finally, let's change our initial value of the integer variable to 0:

```
integer = 0;
```

Now, when we run the program, we see only the following output to the console:

```
Answer: 0! is 1
```

This result is correct because according to the mathematical definition of a factorial, 0! is equal to 1. However, look closely at the console output. The statements in the body of the for statement were *not* executed, not even once. This is because when the value of integer is 0, the condition in the heading of the for loop is *already* false:

```
                           ┌─────────────────┐
                           │ 1     <=    0   │
                           └─────────────────┘
    for (int count = 1; count <= integer; count = count + 1)
    {
        factorial = factorial * count;
        println(count + "! = " + factorial);
    }
```

Thus, in this case, the entire body of the `for` statement is skipped. However, because we initialized the value of the `factorial` variable to 1, we still obtain the correct result when the answer is displayed by the last statement in our program.

Factorials grow large very quickly! We can go up to 16 as the value of our number before we overflow the capacity of the `int` variable type and negative integers begin to be displayed in the console.* Processing also provides the **long** primitive data type for storing larger integers. If we change the declaration of the factorial variable to type `long` as follows, then we can calculate up to the factorial of 20 before overflow occurs:

```
int integer;
integer = 5;

long factorial;
factorial = 1;

for (int count = 1; count <= integer; count = count + 1)
{
  factorial = factorial * count;
  println(count + "! = " + factorial);
}

println("Answer: " + integer + "! is " + factorial);
```

Now, when we run this program, we get the following output to the console:

```
1! = 1
    ⋮
20! = 2432902008176640000
Answer: 20! is 2432902008176640000
```

Save and close the `Factorial` program.

Nested `for` Loops

It is possible to *nest* one `for` statement inside another `for` statement. These are known as **nested** `for` statements.

Nested `for` loops will not be used often in this text, so it's alright if you don't understand everything that is introduced in the following examples. Still, it's good to have a general idea of the kind of effect that nested `for` statements can produce, and these examples are intended to help give this to you.

Select **File** > **New** to start a new program. Save this program as **NestedSeats**.

---

* Negative numbers result because of the way that negative numbers are stored by the computer using binary representation.

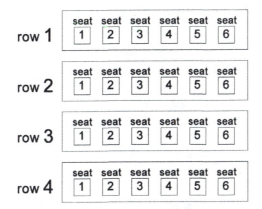

As an illustration of the basic logic of two nested for statements, let's consider an example. Suppose there is a classroom with four rows, numbered 1–4. In each row, there are six seats, numbered 1–6.

Now, suppose we want to write a for statement to count through and display the *seat* numbers that appear in any row. In order to accomplish this, we can write

```
for (int seat = 1; seat <= 6; seat = seat + 1)
{
  print(" Seat " + seat);
}
```

Let's also add a call to the println() function to identify the end of the row:

```
for (int seat = 1; seat <= 6; seat = seat + 1)
{
  print(" Seat " + seat);
}
println(" -- end of row");
```

When we run this program, we see the following output in the console:

```
Seat 1 Seat 2 Seat 3 Seat 4 Seat 5 Seat 6 -- end of row
```

Now, suppose we would instead like to have a for statement to count through and display the *row* numbers. For this, we can write

```
for (int row = 1; row <= 5; row = row + 1)
{
  println("Row: " + row);
}
```

When we run this program, we now see the following output to the console:

```
Row: 1
Row: 2
Row: 3
Row: 4
Row: 5
```

Now, suppose we wish to count through *all* the seats, row by row. The order of this counting would be

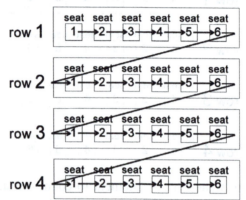

To achieve this order of counting in our program, let's try putting the code for the *seat*-counting `for` statement *inside* the *row*-counting `for` statement:

```
for (int row = 1; row <= 5; row = row + 1)
{
  println("Row: " + row);

  for (int seat = 1; seat <= 6; seat = seat + 1)
  {
    print(" Seat " + seat);
  }
  println(" -- end of row");

}
```

Now, when we run this program, we see the following output to the console:

```
Row: 1
 Seat 1 Seat 2 Seat 3 Seat 4 Seat 5 Seat 6 -- end of row
Row: 2
 Seat 1 Seat 2 Seat 3 Seat 4 Seat 5 Seat 6 -- end of row
Row: 3
 Seat 1 Seat 2 Seat 3 Seat 4 Seat 5 Seat 6 -- end of row
Row: 4
 Seat 1 Seat 2 Seat 3 Seat 4 Seat 5 Seat 6 -- end of row
Row: 5
 Seat 1 Seat 2 Seat 3 Seat 4 Seat 5 Seat 6 -- end of row
```

As you can see, these nested `for` statements have the effect of counting through each of the seats in the classroom, row by row. By nesting one `for` statement inside the other, the *inner* `for` loop performs its *entire* range of counting through the seats each time the *outer* `for` loop advances *one* row ahead in its counting.

Save and close the `NestedSeats` program.

Nested `for` Statements and Pixels: A Graphical Example

Nested `for` statements are somewhat complicated. However, they can also enable us to produce some interesting graphical effects in Processing. For example, nested `for` statements can be very useful when there is some operation that we would like to perform throughout the canvas. To illustrate this, we can manipulate the entire canvas, pixel by pixel, using two nested `for` statements. Select **File > New** to start a new program. Save this program as **NestedPixels**.

We'll use a 300-pixel by 200-pixel canvas:

```
size(300, 200);
```

We can count through each pixel in the canvas in the same way that we did with our classroom seat example. The following `for` statement will count through each pixel *column* of the canvas:

```
for (int column = 0; column < width; column = column + 1)
{

}
```

And the following `for` statement will count through each pixel *row* of the canvas:

```
for (int row = 0; row < height; row = row + 1)
{

}
```

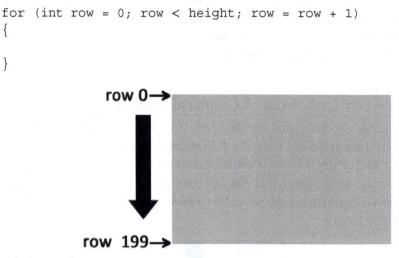

Now, let's *nest* these two `for` statements:

```
size(300, 200);

for (int row = 0; row < height; row = row + 1)
{
  for (int column = 0; column < width; column = column + 1)
  {

  }
}
```

Remember, nesting the two `for` statements in our classroom seat example caused Processing to count through each seat, row by row. In our current program, a similar nesting of `for` statements will cause Processing to count through each pixel *column* number for each pixel *row*. In other words, Processing will count through *every* pixel location on the canvas, pixel by pixel—column by column, row by row.

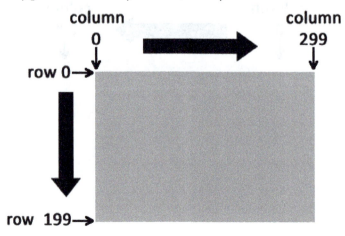

Now, suppose we would like to set each pixel to a random color. We can do this by placing the following two statements inside our nested for statements:

```
size(300, 200);

for (int row = 0; row < height; row = row + 1)
{
  for (int column = 0; column < width; column = column + 1)
  {
    stroke(random(256), random(256), random(256));
    point(column, row);
  }
}
```

When we run this program, we see that Processing does indeed count through each pixel on the canvas, choosing at random a color for each pixel:

If we would instead like to fill the canvas with little "tiles" that are 5 pixels by 5 pixels, we can modify the program as follows:

```
size(300, 200);

for (int row = 0; row < height; row = row + 5)
{
  for (int column = 0; column < width; column = column + 5)
  {
   fill(random(256), random(256), random(256));
   rect(column, row, 5, 5);
  }
}
```

Now, when we run our program, Processing counts through the column numbers in increments of 5 and through the row numbers in increments of 5, drawing 5-pixel by 5-pixel squares of random colors as it makes its way through the canvas. This produces the following "tiling" effect:

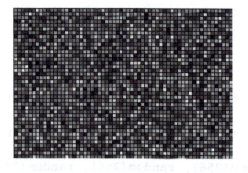

Next, let's try choosing a larger step and square size, 20 pixels, and turn off stroking:

```
size(300, 200);

for (int row = 0; row < height; row = row + 20)
{
  for (int column = 0; column < width; column = column + 20)
  {
    noStroke();
    fill(random(256), random(256), random(256));
    rect(column, row, 20, 20);
  }
}
```

The tiling now looks something like a quilt:

We see this result because Processing is now counting through the column numbers in increments of 20 and through the row numbers in increments of 20, drawing 20-pixel by 20-pixel squares of random colors as it makes its way through the canvas:

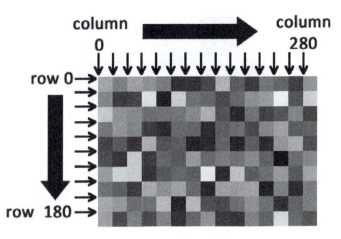

In fact, if we substitute a variable for our step size in place of the four specific values we are currently using, then we can allow the user to enter the tile size:

```
import static javax.swing.JOptionPane.*;

size(300, 200);

String input;
input = showInputDialog("Enter tile size:");

int tile;
tile = int(input);

for (int row = 0; row < height; row = row + tile)
{
  for (int column = 0; column < width; column = column + tile)
  {
    noStroke();
    fill(random(256), random(256), random(256));
    rect(column, row, tile, tile);
  }
}
```

Now, when we run our program, the user can enter a tile size:

Processing then fills the canvas with randomly colored squares of the specified size:

Processing also has the ability to work with photos, and this technique of using nested for statements to step through the canvas, pixel by pixel, can be used to produce some fun and powerful photo manipulation.*

SUMMARY

In this chapter, we learned about using the for statement to create counting loops.

- We learned how for loops are useful for counting through sequences of numbers.

- We learned how for loops are useful for repeating a set of actions a certain number of times.

- We learned how a loop counter variable can be used to count upward or downward.

- We learned how a loop counter variable can be increased (incremented) or decreased (decremented) by 1 or another amount.

- We saw how one for statement can be nested inside another for statement.

Exercises

1) Write a for statement to produce the following output to the console:

```
1
2
3
4
5
6
```

---

* If you're interested in trying photo manipulation with nested for statements, you can explore the earth image example in the online "Two-Dimensional Arrays" chapter. Nested for statements are powerful but somewhat tricky to use. They will not appear in the remainder of this book.

2) Write a for statement to produce the following output to the console:

```
2
3
4
5
```

3) Write a for statement to produce the following output to the console:

```
2
4
6
8
10
12
```

4) Write a for statement to produce the following output to the console:

```
5
10
15
20
25
30
```

5) Write a for statement to produce the following output to the console:

```
Processing
Processing
Processing
Processing
Processing
Processing
Processing
Processing
Processing
Processing
```

6) Write a for statement to produce the following output to the console:

```
line: 1
line: 2
line: 3
line: 4
line: 5
line: 6
line: 7
line: 8
line: 9
```

7) Write a `for` statement to produce the following output to the console:

```
50
45
40
35
30
25
20
15
10
5
0
```

8) Write a `for` statement to output the squares of the integers 1–6 in the console:

```
1
4
9
16
25
36
```

9) Write a `for` statement to produce the following output to the console:

```
0
20
40
60
80
100
120
140
160
180
200
```

10) Modify the previous program by doing the following:

  a) Set the canvas size to 200 pixels by 200 pixels.

  b) Using the `width` and `height` variables, add a call to the `line()` function inside the body of the `for` statement so that the following line segments are drawn on the canvas:

11) Modify the call to the `line()` function in the previous program so that the following line segments are drawn on the canvas:

12) Modify the call to the `line()` function in the previous program so that the following line segments are drawn on the canvas:

13) Modify the call to the line() function in the previous program so that it instead draws the following horizontal lines on the canvas:

14) Modify the call to the line() function in the previous program so that it instead draws the following vertical line segments on the canvas:

15) Add a second call to the line() function inside the body of the for statement in the previous program so that both horizontal and vertical lines are drawn on the canvas:

16) Write a for statement to produce the following output to the console:

```
30
60
90
120
150
180
```

17) Modify the previous program by doing the following:

a) Set the canvas size to 200 pixels by 200 pixels.

b) Add a call to the ellipse() function inside the for statement so that the following circles with diameters of 20 pixels are drawn on the canvas:

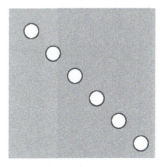

18) Write a for statement to produce the following output to the console:

```
150
125
100
75
50
```

19) Modify the previous program by doing the following:

a) Set the canvas size to 200 pixels by 200 pixels.

b) Add a call to the rect() function inside the for statement so that the following squares are drawn on the canvas:

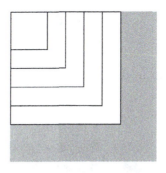

20) Write a program to do the following:

    a) Set the canvas size to 200 pixels by 200 pixels.

    b) Then, use a `for` statement to draw five circles with diameters of 20 pixels at random locations on the canvas.

    Sample output

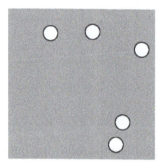

21) Modify the previous program by inserting a call to the `fill()` function inside the `for` statement so that each circle is drawn with a random fill color.

    Sample run

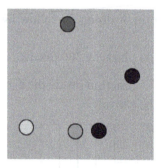

22) Modify the `for` statement in the previous program so that 100 circles are drawn instead.

    Sample run

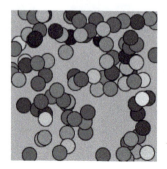

23) Write a program that uses a for loop to calculate the sum of all the positive integers up to a certain number.* For example, if the number is 5, the following output would be produced:

```
1
2
3
4
5
Total: 15
```

24) Modify the previous program so that the user inputs the number.

---

* You can test your results using the Gauss formula by adding the following statement to your program: println("Gauss:"
+ number * (number + 1) / 2);

CHAPTER 7

Creating `void` Functions

Thus far, we have been able to *use* many `void` and non-`void` functions that are built into Processing. However, it is also possible for us to *create* our own `void` and non-`void` functions. Creating our own functions can provide us with a number of benefits, including the following:

- Functions can make our code *easier to read*.

- Functions allow us to *reuse a* sequence of statements that, together, comprise some desirable overall *action*.

In this chapter, we will learn how to create what are known as `void` functions.

`void` Functions

We all have various *routines* that we perform as part of our lives. For example, a parent might give a child the instruction,

"Get ready for school."

Both the parent and child understand that this general instruction—"Get ready for school"—actually translates into a whole sequence of more specific instructions, such as

1) "Get out of bed."

2) "Take a shower."

3) "Brush your hair."

4) "Get dressed."

5) "Eat breakfast."

6) "Brush your teeth."

7) "Get your backpack."

8) "Get in the car."

Because the individual steps of this named routine are predefined, the parent can now simply give the child the general instruction, "Get ready for school." The child will then know to perform the corresponding sequence of individual activities that, together, comprise the desired overall action that the general instruction describes: getting ready for school.

In Processing, a **void function** is something similar: it is a sequence of statements that has been given a name and will be performed whenever this function is called by name.* Performing this sequence of statements comprises the overall action that is described by the function's name.

As we have seen, Processing has many *built-in* void functions: for example, println(). Behind the scenes, println() is a void function that is predefined in Processing to perform a sequence of steps that, together, comprise the desired overall action described by the function's name: displaying a line of text in the console.

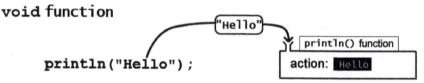

To call to one of Processing's void functions, we simply give the *name* of the function followed by any *argument(s)* that we are required to supply to this particular function. For example, we might call the println() function by writing

```
println("Hello");
```

When we run our program and one of Processing's built-in void functions is called, that function always performs the specific overall action it was predefined to perform, using the particular argument(s) we supplied in the function call. In our current example, the println() function will display "Hello" on a line of text in the console.

As we will see, we call any void function that we create ourselves in the same manner. The key difference is, when a function of our own creation is called, it will perform the specific sequence of actions that *we* have defined.

Active Mode: Introducing the setup() Function

Select **File > New** to start a new program. Save this program as **Active**.

---

* A void function is also known as a *procedure* or *subroutine*.

Until now, we have been writing our programs in what is known as Processing's *static* mode. However, in order to create and use our own functions, we need to work in what is known as Processing's *active* mode. Our work in active mode will always involve defining the special **setup()** function. To define the setup() function, we can start by simply writing

```
void setup()
{

}
```

Then, inside the curly braces, we simply place the sequence of statements that we would like to have performed when the setup() function is called. For example, to have the setup() function display a greeting in the console, we might add

```
void setup()
{
  println("Hello!");
  println("How are you?");
}
```

These steps for *defining* the special setup() function is very similar to the way that we will define our own void functions. However, the setup() function is special in the way it is *called*. As we have seen, we normally write an explicit call to a function whenever we want that function to perform its predefined action. However, in active mode, whenever we run our program, Processing will always begin by automatically making a single call to the setup() function. In other words, in active mode, when we run our programs, they always automatically begin with the setup() function. For this reason, in active mode, the setup() function is where we want to put the statements that make up our main program.

This also means that we can convert any one of the *static* mode programs we have written thus far to an *active* mode program simply by placing its statements inside the setup() function. When we run this new active mode version of the program, it will perform exactly the same as the static mode version. For example, consider our simplest version of the dice-rolling game:

```
int roll;
roll = int( random(1, 7) );
println("Roll: " + roll);
```

This program produces a random value from 1 to 6 and displays it in the console. For example:

```
Roll: 5
```

This program was written in Processing's *static* mode. In order to convert this program to *active* mode, we can simply place all its statements inside the curly braces at the beginning and end of the setup() function:

```
void setup()
{
  int roll;
  roll = int( random(1, 7) );
  println("Roll: " + roll);
}
```

When we run this program, Processing starts by *automatically* calling the setup() function *once*. The statements in the setup() function are then performed in order. Afterward, the performance of the setup() function is complete. Thus, this *active* mode version of our program works just as the *static* mode version did, and we see exactly the same kind of output to the console:

```
Roll: 6
```

Delete the statements currently inside the curly braces of the setup() function. Then, save the Active program. We will continue working with it in the next section.

A Closer Look at the setup() Function

Return to the Active program. As we saw in the preceding section, we begin the definition of the setup() function by writing the following:

```
void setup()
{

}
```

Let's look a little more closely at the code we've entered. The first line of the function definition is called the function **header**.

$$\boxed{\texttt{void setup()}} \leftarrow \text{header}$$
$$\{$$

$$\}$$

A function's header always consists of *three* items of information:

1) The *type of value* that is returned by the function once it has completed its defined actions.

$$\text{type of value returned}$$
$$\downarrow$$
$$\boxed{\texttt{void}}\, \texttt{setup()}$$
$$\{$$

$$\}$$

The setup() function does not return a resulting value to us when it is finished. In Processing terminology, this means that it is a void function. Thus, the return type is specified as void in the function header.

2) The function's *name*.

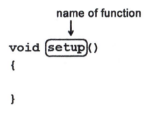

The function name is the next item added to the function header. The name of this function is "setup." The first letter of a function name is typically a *lowercase* letter.

3) The function's required *parameters*, enclosed in a pair of parentheses.

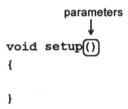

As we will see later in this chapter, parameters are *variables* that store any values that need to be sent to this function as arguments whenever it is called. The setup() function does not have any required parameter variables, so we simply add an empty pair of parentheses to the function header.

Thus, the three items of information that comprise the *header* of a function definition are

1) The *type of value returned*

2) The *name* of the function

3) The *parameters* required

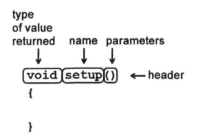

After the function header comes the function **body**. This portion of the function definition is enclosed between a pair of curly braces:

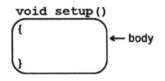

Between these braces, we place any statements that we want to be performed when the setup() function is called. For example, let's have the setup () function display information in the console announcing when the performance of the statements in the body of this function is beginning and ending:

```
void setup()
{
  println("Beginning the setup() function...");
  println("The setup() function has finished.");
}
```

As we have learned, the setup() function is special. When we run an active mode program containing the setup() function, Processing will begin by *automatically* calling this function, just *once*. Thus, when we run our current program, the setup() function is immediately performed a single time, and we see the following output to the console:

```
Beginning the setup() function...
The setup() function has finished.
```

Save the Active program. We will continue working with it in the next section.

Creating Our Own void Function

Return to the Active program. Let's try adding a *new* void function of our own creation to this program, a function that is *not* already built into Processing. We define such a function for an active mode program in very much the same way we defined the setup() function: by defining the function *header* and the function *body*.

First, we start with the three required parts of the *header* of our function:

1) The *type of value that will be returned.*

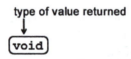

In this chapter, we're focusing on void functions, which are functions that do not return a value. Thus, the return type we specify here is **void**.

2) The *name* of the function.

name of function

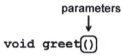

Let's use "greet" as the name of this function.

3) The *parameters* required by the function, enclosed in a pair of parentheses.

parameters

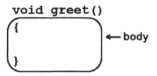

To keep our first function simple, we won't have any required parameters. Thus, we simply add an empty pair of parentheses.

In addition to defining the header of our function, we need to define the **body** of our function. We can start this definition by placing statements between a pair of curly braces below the function header.

void greet()

```
{

}              ← body
```

Let's have this greet() function issue a greeting: "Hello!" But let's also have the greet () function display in the console announcements that the performance of the statements in the body of this function is beginning and ending. As a result, the body of our new greet() function is

```
void greet()
{
  println(" Beginning the greet() function...");

  println(" Hello!");

  println(" The greet() function has finished.");
}
```

And our complete program is now

```
void setup()
{
  println("Beginning the setup() function...");
  println("The setup() function has finished.");
}
```

```
void greet()
{
  println(" Beginning the greet() function...");
  println(" Hello!");
  println(" The greet() function has finished.");
}
```

However, when we run this program, we see the same console output as before:

```
Beginning the setup() function...
The setup() function has finished.
```

Why didn't we see the announcements or the greeting from the greet() function? In active mode, the setup() function is *automatically* called *once* by Processing as soon as we start the run of our program. However, unlike the setup() function, any function that *we* create ourselves must be *explicitly* called in order for the statements it contains to be performed. Thus, we need to add an explicit call to our greet() function inside the setup() function.

How do we create an explicit call to our greet() function? As we have learned, a call to a void function consists of the function *name*, followed by a list of any necessary *arguments* enclosed in a pair of parentheses followed by a semicolon. Thus, to create a call to our new greet() function, which is a void function, we can simply add the following statement inside the setup() function:

```
void setup()
{
  println("Beginning the setup() function...");

  greet();

  println("The setup() function has finished.");
}

void greet()
{
  println(" Beginning the greet() function...");

  println(" Hello!");

  println(" The greet() function has finished.");
}
```

Now, when we run this program, our `greet()` function is called as part of the automatic performance of the `setup()` function. Thus, we see the following output to the console:

```
Beginning the setup() function...
Beginning the greet() function...
Hello!
The greet() function has finished.
The setup() function has finished.
```

This particular output results because of the *order* in which the statements are executed—also known as the **flow of execution**. Let's examine our program's flow of execution, step by step. When we run our program, Processing immediately calls the `setup()` function automatically. As a result, the statements in the body of the `setup()` function are performed in sequential order. Thus, the first statement of the `setup()` function is performed:

```
void setup()
{
  println("Beginning the setup() function...");

  greet();

  println("The setup() function has finished.");
}
```

And the following console output results:

```
Beginning the setup() function...
```

The next statement contains, a call to the `greet()` function:

```
void setup()
{
  println("Beginning the setup() function...");

  greet();

  println("The setup() function has finished.");
}
```

As a result of this function call, the flow of execution now jumps to the `greet()` function:

```
void setup()
{
   println("Beginning the setup() function...");

   greet();

   println("The setup() function has finished.");
}

void greet()
{
   println("  Beginning the greet() function...");

   println("  Hello!");

   println("  The greet() function has finished.");
}
```

The statements in the body of the greet() function are now performed in the order in which they are listed, starting with the first statement:

```
void greet()
{
  println(" Beginning the greet() function...");

  println(" Hello!");

  println(" The greet() function has finished.");
}
```

This statement produces another line of console output:

```
Beginning the greet() function...
```

Next, the second statement in the greet() function is performed:

```
void greet()
{
  println(" Beginning the greet() function...");

  println(" Hello!");

  println(" The greet() function has finished.");
}
```

This statement produces another line of console output:

Hello!

Finally, the last statement of the greet() function is performed:

```
void greet()
{
  println(" Beginning the greet() function...");

  println(" Hello!");

  println(" The greet() function has finished.");
}
```

This statement also produces a line of console output:

The greet() function has finished.

The performance of the statements in the body of the greet () function is now complete. What happens next? When the performance of the statements in the body of a function has been completed, the flow of execution always *returns to the point from which the function was called*. Thus, once Processing reaches the end of the body of the greet() function, the flow of execution then returns to the exact point in the setup() function where the greet() function was called.

```
void setup()
{
  println("Beginning the setup() function...");

  greet();  ←

  println("The setup() function has finished.");
}

void greet()
{
  println("  Beginning the greet() function...");

  println("  Hello!");

  println("  The greet() function has finished.");
}
```

Flow of execution returns to the point from which the function was called.

The performance of the statements in the body of the setup() function now continues, starting at this return point. Thus, the statement immediately following the call to the greet() function is performed next:

```
void setup()
{
  println("Beginning the setup() function...");

  greet();

  println("The setup() function has finished.");
}
```

This statement produces the final line of console output:

```
The setup() function has finished.
```

Because all the statements in the body of the setup() function have been performed, the execution of setup() is now complete. Thus, the execution of our program is finished. The total output to the console produced by our program is:

```
Beginning the setup() function...
Beginning the greet() function...
Hello!
The greet() function has finished.
The setup() function has finished.
```

We have now defined a void function of our own: greet(). We have successfully called this function from the setup() function. We have also learned a bit about the flow of execution that results from calling a function.

Save and close the Active program.

Graphical Example of a void Function

Select **File > New** to start a new program. Save this program as **BalloonDraw**.

Suppose we have the following setup() function that generates a canvas that is 200 pixels by 200 pixels:

```
void setup()
{
  size(200, 200);
}
```

Let's define and use a function named balloon() that will draw a red balloon on this canvas at a random pixel location:

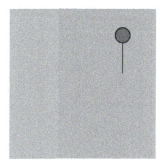

We'll start the definition of our balloon() function by defining the function *header*, which, as we know, always requires *three* items of information:

1) The *type of value returned*.

   We do not need this function to return a value, so the return type is void.

   void

2) The function *name*.

   The function name will be "balloon."

   void **balloon**

3) The *parameters* required.

   We'll define this version of the function so that it doesn't need us to supply any information in order for the balloon to be drawn. Thus, no parameters are needed, so we simply add an empty pair of parentheses:

   void balloon()

The definition of the function header is now complete:

```
void balloon()
```

We now define the function *body*, enclosed between a pair of curly braces:

```
void balloon()
{

}
```

The balloon drawn by the `balloon()` function will consist of simply a circle and a line segment. Let's define and initialize two variables to store (1) the length of the line segment (the balloon's string) and (2) the diameter of the balloon itself. The values of these two variables won't change during our program, but using variables will make our code more readable than if we use "magic numbers" (unexplained specific values). Also, to keep our code a little more concise in this example, let's take advantage of the option of declaring and initializing each of these variables in a single statement:

```
void balloon()
{
  int lineLength = 50;
  int diameter = 20;
}
```

For now, we'll pick both the pixel *column* location and the pixel *row* location of the center of the balloon at random. These pixel column location and pixel row locations must be stored in variables, because, as we'll see, we will need to use this pair of variables *twice*: first to draw the circle (balloon) and then to draw the line segment (the balloon's string). Here, too, let's declare each of these two variables and assign each one a random value in a single statement:

```
void balloon()
{
  int lineLength = 50;
  int diameter = 20;

  int column = int( random(width) );
  int row = int( random(height) );
}
```

All four of these variables (`lineLength`, `diameter`, `column`, and `row`) are examples of what are known as **local** variables. The **scope** (range of use) of a local variable is limited to the single statement or block of statements (defined by a pair of curly braces) in which that variable is declared. Thus, in this particular case, none of these local variables can be used outside of the `balloon()` function. However, this does not pose any problem for us, because we will only be using these variables for the very specific task that we are defining this function to perform: drawing a balloon at a random pixel location.

We're ready to add to our `balloon()` function the statements necessary to draw the balloon. Because we want the balloon's "body" to be drawn *on top of* the balloon's string, we'll draw the balloon's string *first*. For this, we will start our line segment at the chosen pixel location stored in our two variables, `column` and `row`. The other end of the line segment will be at the pixel location that is in the same column but is 50 rows down, the value of our `lineLength` variable. Thus, the location of the second pixel of this line segment can be described as column location `column` and row location `row + lineLength`. We can visualize this as

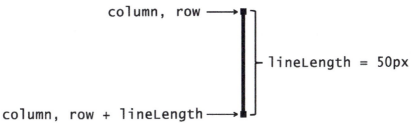

Thus, we will add the following statement to our `balloon()` function:

```
void balloon()
{
  int lineLength = 50;
  int diameter = 20;

  int column = int( random(width) );
  int row = int( random(height) );

  line(column, row, column, row + lineLength);
}
```

Next, we'll draw the balloon's "body." For this, we will set the fill color to red. Then, we will draw a circle at the randomly chosen pixel location with a diameter equal to the value of our `diameter` variable. For this, we add the following two statements to our `balloon()` function:

```
void balloon()
{
  int lineLength = 50;
  int diameter = 20;

  int column = int( random(width) );
  int row = int( random(height) );

  line(column, row, column, row + lineLength);
  fill(255, 0, 0);
  ellipse(column, row, diameter, diameter);
}
```

The circle representing the balloon's body will *overlap* part of the line segment that represents the balloon's string:

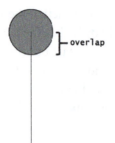

However, this overlap will not be visible because the fill color of the circle is fully *opaque* by default.

Our program is now

```
void setup()
{
  size(200, 200);
}

void balloon()
{
  int lineLength = 50;
  int diameter = 20;

  int column = int( random(width) );
  int row = int( random(height) );

  line(column, row, column, row + lineLength);
  fill(255, 0, 0);
  ellipse(column, row, diameter, diameter);
}
```

However, when we run this program, we see no balloon drawn on the canvas:

We see that a 200-pixel by 200-pixel canvas was rendered, so we know that the setup() function was performed. Indeed, as usual, the setup() function was *automatically* called a single time by Processing when we ran our program.

However, our `balloon()` function was *not* performed. This is because, as we have learned, any function that we define ourselves is *not* called automatically. Rather, we must *explicitly* call our function in order for it to be performed. Thus, we simply need to insert an explicit call to our `balloon()` function inside the body of the `setup()` function:

```
void setup()
{
  size(200, 200);

  balloon();

}
void balloon()
{
  int lineLength = 50;
  int diameter = 20;

  int column = int( random(width) );
  int row = int( random(height) );

  line(column, row, column, row + lineLength);
  fill(255, 0, 0);
  ellipse(column, row, diameter, diameter);
}
```

Now, when we run our program, a balloon is indeed drawn at a randomly chosen location on the canvas:

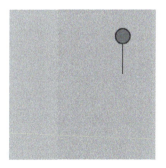

Let's once again trace the flow of execution of our function. When we run our program, Processing starts by automatically calling the `setup()` function, and the statements in the body of that function are performed. Thus, the call to the `size()` function is performed and a canvas that is 200 pixels by 200 pixels is generated:

```
void setup()
{
  size(200, 200);

  balloon();

}
```

The next statement performed contains a call to the `balloon()` function:

```
void setup()
{
  size(200, 200);

  balloon();

}
```

Here, the flow of execution jumps to the `balloon()` function. We can visualize this as

```
void setup()
{
  size(200, 200);

  balloon();


  }

▶void balloon()
  {
    int lineLength = 50;
    int diameter = 20;

    int column = int( random(width) );
    int row = int( random(height) );

    line(column, row, column, row + lineLength);
    fill(255, 0, 0);
    ellipse(column, row, diameter, diameter);
  }
```

Next, the statements in the body of the `balloon()` function are performed. As a result, a balloon is drawn at a randomly chosen pixel location:

```
void balloon()
{
  int lineLength = 50;
  int diameter = 20;

  int column = int( random(width) );
  int row = int( random(height) );

  line(column, row, column, row + lineLength);
  fill(255, 0, 0);
  ellipse(column, row, diameter, diameter);
}
```

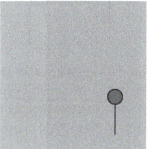

Once the end of the body of the balloon() function has been reached, the flow of execution then *returns* to the setup() function, to the exact point at which the balloon() function was called. We can visualize this return as

```
void setup()
{
  size(200, 200);

  balloon(); ←

}

void balloon()
{                                        Flow of execution
  int lineLength = 50;                   returns to the point
  int diameter = 20;                     from which the
                                         function was called.
  int column = int( random(width) );
  int row = int( random(height) );

  line(column, row, column, row + lineLength);
  fill(255, 0, 0);
  ellipse(column, row, diameter, diameter);
}
```

Because the call to the balloon () function is the last statement in the setup() function, the execution of this program is now complete.*

Save this BalloonDraw program. We will return to it later in this chapter.

Reusing a Function

Now, what if we would like to have *multiple* balloons drawn on the canvas? This is one of the benefits of a function: it is *reusable*.

Return to the BalloonDraw program. Select **File > Save As** and resave this program as **Balloons**.

Because we have placed the balloon-drawing statements inside the balloon () function, we can *repeat* the performance of these statements simply by calling this function again. We write one explicit call to the balloon() function one time for each balloon that we would like to have drawn on the canvas. For example, if we would like to draw three balloons, we can write

```
void setup()
{
  size(200, 200);

  balloon();

  balloon();

  balloon();
}

void balloon()
{
  int lineLength = 50;
  int diameter = 20;

  int column = int( random(width) );
  int row = int( random(height) );

  line(column, row, column, row + lineLength);
  fill(255, 0, 0);
  ellipse(column, row, diameter, diameter);
}
```

When we run this program, we see that *each* of the three times the balloon () function is called

---

* In this active mode program, there is no draw() function, which Processing would have called automatically after the completion of the setup() function.

1) The flow of execution jumps to the balloon() function.

2) The balloon-drawing statements in the body of the balloon() function are performed.

3) The flow of execution returns to the point from which the balloon() function was called.

The first call to the balloon() function produces the first balloon:

```
void setup()
{
  size(200, 200);

  balloon();

  balloon();

  balloon();

}

void balloon()
{
  int lineLength = 50;
  int diameter = 20;

  int column = int( random(width) );
  int row = int( random(height) );

  line(column, row, column, row + lineLength);
  fill(255, 0, 0);
  ellipse(column, row, diameter, diameter);
}
```

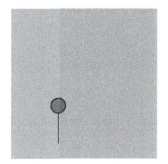

The second call to the balloon() function produces a second balloon:

```
void setup()
{
  size(200, 200);

  balloon();

  balloon();

  balloon();

}

void balloon()
{
  int lineLength = 50;
  int diameter = 20;

  int column = int( random(width) );
  int row = int( random(height) );

  line(column, row, column, row + lineLength);
  fill(255, 0, 0);
  ellipse(column, row, diameter, diameter);
}
```

And the third call to the balloon() function produces a third balloon:

```
void setup()
{
  size(200, 200);

  balloon();

  balloon();

  balloon();

}

void balloon()
{
  int lineLength = 50;
  int diameter = 20;

  int column = int( random(width) );
  int row = int( random(height) );

  line(column, row, column, row + lineLength);
  fill(255, 0, 0);
  ellipse(column, row, diameter, diameter);
}
```

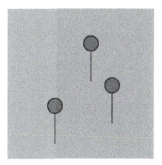

As we can see, because the setup() function contains three calls to the balloon() function, three randomly located balloons are drawn on the canvas each time we run this program.

Another way to generate multiple balloons is by enclosing a single call to the balloon() function inside a for statement. We can then select the number of balloons that will be drawn (e.g., five) simply by setting the loop condition in the heading of the for statement:

```
void setup()
{
  size(200, 200);

  for (int count = 1; count <= 5; count = count + 1)
  {
    balloon();
  }

}

void balloon()
{
  int lineLength = 50;
  int diameter = 20;

  int column = int( random(width) );
  int row = int( random(height) );

  line(column, row, column, row + lineLength);
  fill(255, 0, 0);
  ellipse(column, row, diameter, diameter);
}
```

Now, when we run our program, each time the body of the for loop is performed (in this case, five times), the flow of execution jumps to and returns from the balloon() function, and a randomly located balloon is drawn on the canvas each time.

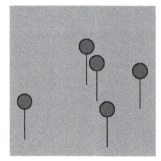

Save and close the Balloons program.

Function Parameters and Arguments

So far in this chapter, each function we have defined has featured an *empty* pair of parentheses in both the header of the function definition

```
void balloon()
```

and in any call to the function

```
balloon();
```

However, a function may also be defined such that it requires one or more items of information to be *passed* to it each time that function is called. As we have learned, such required items are known as **parameters**, and the items that we supply to meet these requirements when we call such a function are known as **arguments**.

Example: Adding a Parameter to a Function

Let's return to our `BalloonDraw` program. Currently, the diameter of the balloon is specified inside the `balloon()` function using an assignment statement and a local variable:

```
void setup()
{
  size(200, 200);

  balloon();

}

void balloon()

{
  int lineLength = 50;
  int diameter = 20;

  int column = int( random(width) );
  int row = int( random(height) );

  line(column, row, column, row + lineLength);
  fill(255, 0, 0);
  ellipse(column, row, diameter, diameter);
}
```

Instead of having the circle's diameter fixed at 20 pixels, let's modify our `balloon()` function so it will draw a balloon with whatever diameter we specify. For this, we need to define a *parameter* variable for the `balloon()` function so that we can pass our desired diameter to the `balloon()` function as an *argument* whenever we call this function.

As we know, the list of parameters, enclosed in a pair of parentheses, is one of the items of information we specify in a function's header. Currently, the `balloon()` function has no parameters declared in its header:

```
void balloon()
```

Declaring a parameter variable is similar to defining any other variable: we specify first the *type* of the variable and then its *name*. However, when declaring a parameter variable, we put this declaration inside the function *header*.

Thus, to add a diameter parameter to the header of the `balloon ()` function, we write

```
void balloon(int diameter)
```

We now have a parameter variable named `diameter` that will serve as a place to store a copy of the diameter argument value that will need to be passed to the `balloon()` function whenever we call this function.

A parameter variable is a *local* variable. Thus, the *scope* of a parameter variable is such that we may only use it within the function itself. This means that we can only use this parameter variable `diameter` *inside* the `balloon()` function. Notice that underlining has now appeared in our statement that declared and initialized the `diameter` variable:

```
int diameter = 20;
```

And an error message has appeared in the Message Area:

Duplicate local variable diameter

As this error message indicates, now have *two* local variables named `diameter` declared inside the same function:

```
void balloon(int diameter)
{

    int lineLength = 50;
    int diameter = 20;

    int column = int( random(width) );
    int row = int( random(height) );

    line(column, row, column, row + lineLength);
    fill(255, 0, 0);
    ellipse(column, row, diameter, diameter);
}
```

We can *delete* the statement in the *body* of the `balloon()` function that declares and initializes a `diameter` variable:

```
int diameter = 20;
```

We no longer need this statement because, instead of assigning our `diameter` variable the same specific value each time, 20, we are now going to be supplying a value for the `diameter` variable as an *argument* each time the `balloon()` function is called. Thus, our program is now

```
void balloon(int diameter)
{
  int lineLength = 50;

  int column = int( random(width) );
  int row = int( random(height) );

  line(column, row, column, row + lineLength);
  fill(255, 0, 0);
  ellipse(column, row, diameter, diameter);
}
```

Save the BalloonDraw program. We will continue using it in the next section.

Supplying an Argument to a Function

Return to the BalloonDraw program. As you may have noticed, our call to the balloon() function inside the setup() function is now underlined:

```
void setup()
{

  size(200, 200);

  balloon();

}
```

And the corresponding error message displayed in the Message Area reads:

The function balloon() expects parameters like "balloon(int)"

We are now seeing this error message because we have added a parameter variable, diameter, to our balloon() function. This means that, from now on, whenever we call the balloon() function, we must supply an *argument* value to be assigned to this parameter variable. The list of *arguments* that we supply in a function call must always *match* the list of *parameters* in the function definition, in three ways:

1) The *number* of arguments must be the same as the number of parameters.

2) The *order* of the arguments must match the order of the parameters.

3) The *type* of each argument must match the type of the corresponding parameter.*

---

* As usual, if an int value is supplied as an argument for a float parameter variable, the int value will be automatically converted to a float value by Processing, so no error results.

Currently, there is a *mismatch* between the argument list in our call of the balloon()
function and the parameter list in the header of our balloon() function definition.
Specifically, the argument list in the function call is *empty*, but the parameter list in the
function definition contains the declaration of the diameter variable:

```
void setup()
{
   size(200, 200);

   balloon();

}
                    mismatch
void balloon(int diameter)
{
   int lineLength = 50;

   int column = int( random(width) );
   int row = int( random(height) );

   line(column, row, column, row + lineLength);
   fill(255, 0, 0);
   ellipse(column, row, diameter, diameter);
}
```

We need to change our call to the balloon() function so that our argument list matches
the parameter list. The balloon() function now expects a single int value to be passed
to it as an argument each time the function is called. Thus, we will supply 40 as an argument in our call of the balloon() function:

```
void setup()
{
   size(200, 200);

   balloon(40);

}
```

Once we make this change, the underlining and the error message disappear because the
argument list in our function call now matches the *parameter* variable list defined in the
function header.

```
void setup()
{
   size(200, 200);

   balloon (40);
                      match

}

void balloon (int diameter)
{
   int lineLength = 50;

   int column = int( random(width) );
   int row = int( random(height) );

   line(column, row, column, row + lineLength);
   fill(255, 0, 0);
   ellipse(column, row, diameter, diameter);
}
```

Let's trace the flow of execution that results when we run this program. Processing automatically calls the setup() function, and the statements that it contains are performed in sequential order. The first statement in the setup() function is a call to the size() function, which generates a 200-pixel by 200-pixel canvas:

The next statement in the setup() function is a call to the balloon() function. At this point, the flow of execution jumps to the balloon() function, and a copy of the value that we supplied as an argument—in this case, 40—is also passed along and is stored in the diameter parameter variable. We can visualize this as

```
void setup()
{
  size(200, 200);

  balloon(40);

}                    40

void balloon(int diameter)
{
  int lineLength = 50;

  int column = int( random(width) );
  int row = int( random(height) );

  line(column, row, column, row + lineLength);
  fill(255, 0, 0);
  ellipse(column, row, diameter, diameter);
}
```

The statements in the balloon() function are now performed in sequential order. The last statement in the balloon() function is a call to the ellipse() function that makes use of the value stored in the diameter parameter variable to draw the circle portion of the balloon. We can visualize this as

```
                    40
void balloon(int diameter)
{
  int lineLength = 50;

  int column = int( random(width) );
  int row = int( random(height) );

  line(column, row, column, row + lineLength);
  fill(255, 0, 0);
              40        40
  ellipse(column, row, diameter, diameter);
}
```

Thus, the circle that is drawn in a randomly chosen location by the balloon() function has a diameter equal to the argument that we supplied in the function call, 40 pixels:

Now that all the statements in the body of the `balloon()` function have been performed, the flow of execution returns to the point from which this function was called. We can visualize this return as

```
void setup()
{
   size(200, 200);

   balloon(40); ←─────────────────────────────────┐
                                                    │
}                                                   │
                                                    │
void balloon(int diameter)                          │
{                                            Flow of execution
   int lineLength = 50;                      returns to the point
                                             from which the
   int column = int( random(width) );        function was called.
   int row = int( random(height) );          │
                                              │
   line(column, row, column, row + lineLength);
   fill(255, 0, 0);                           │
   ellipse(column, row, diameter, diameter);  │
}─────────────────────────────────────────────┘
```

Because the statements in the body of the `setup()` function have all been performed, the execution of this program is now complete.

By adding a parameter, we made our `balloon()` function more versatile in that we are now able to specify the diameter of the balloon that is drawn.

Save the `BalloonDraw` program. We will continue with it in the next section.

Adding Multiple Parameters to a Function

Suppose that we would also like the `balloon ()` function to draw the balloon at a *specified* pixel location rather than at a randomly chosen one. For this, we need to add two more parameter variables to our function header to specify the pixel column and pixel row where we would like the center of the balloon's circle to be. When there are *multiple* parameter variables, they are listed, separated by *commas*, in the function header. In this list, we must specify the *type* of *each* parameter variable, even if some of these parameter variables are of the same type. Thus, in the `balloon()` function's header, we insert the following:

```
void balloon(int column, int row, int diameter)
```

However, once we make these changes, underlining appears in the following statements in our `balloon()` function:

```
int column = int( random(width) );
int row = int( random(height) );
```

If we click on these underlined portions of our code, we see that the corresponding error messages are, respectively,

Duplicate local variable column

and

Duplicate local variable row

Once again, we have the problem of two local variables being declared with the same name in the same function.

```
void balloon(int column, int row, int diameter)
{
  int lineLength = 50;

  int column = int( random(width) );
  int row = int( random(height) );

  line(column, row, column, row + lineLength);
  fill(255, 0, 0);
  ellipse(column, row, diameter, diameter);
}
```

When we call the `balloon()` function, we will now supply a specific pixel location to this function using the new `column` and `row` parameter variables that we have defined in the `balloon()` function *header*:

```
void balloon(int column, int row, int diameter)
```

We should no longer declare these two local variables and assign them random values. We need to delete these two statements in the *body* of the balloon() function:

```
int column = int( random(width) );
int row = int( random(height) );
```

Thus, our program is now

```
void setup()
{
  size(200, 200);

  balloon(40);

}

void balloon(int column, int row, int diameter)
{
  int lineLength = 50;

  line(column, row, column, row + lineLength);
  fill(255, 0, 0);
  ellipse(column, row, diameter, diameter);
}
```

However, underlining has once again appeared in the call to the balloon() function in the setup() function:

```
void setup()
{
  size(200, 200);

  balloon(40);

}
```

The error message corresponding to this underlining reads

The function balloon() expects parameters like "balloon(int, int, int)"

As we have learned, the list of *arguments* that we supply in a function call must always *match* the list of *parameters* in the function definition in *three* ways:

1) The *number* of arguments

2) The *order* of the arguments

3) The *type* of each argument

We currently have a *mismatch* between the *argument* list in our function call and the *parameter* list in the function header:

```
void setup()
{
  size(200, 200);

  balloon(40);

}

void balloon(int column, int row, int diameter)
{
  int lineLength = 50;

  line(column, row, column, row + lineLength);
  fill(255, 0, 0);
  ellipse(column, row, diameter, diameter);
}
```

We can resolve this mismatch simply by specifying *three* int values for the *argument* list in our function call. For example, to draw a circle with its center at pixel column 75 and pixel row 100 with a diameter of 30 pixels, we can write

```
void setup()
{
  size(200, 200);

  balloon(75, 100, 30);

}
```

Notice that, like the parameter variables in the parameter list, the arguments in the argument list are also separated by *commas*. When we make these changes, the underlining in our call to the balloon() function disappears. This is because we no longer have a mismatch between the argument list in our function call and the parameter list in the function header:

```
void setup()
{
   size(200, 200);

   balloon(75, 100, 30);

}
                    match    match  match

void balloon(int column, int row, int diameter)
{
   int lineLength = 50;

   line(column, row, column, row + lineLength);
   fill(255, 0, 0);
   ellipse(column, row, diameter, diameter);
}
```

Notice that the *order* in which the arguments are listed is important, because these values are assigned to the parameter variables in the same order:

1) The first argument, 75, is assigned to the first parameter variable, column.

2) The second argument, 100, is assigned to the second parameter variable, row.

3) The third argument, 30, is assigned to the third parameter variable, diameter.

Let's once again trace the flow of execution that results when we run this program. Processing automatically calls the setup() function, once, and the statements in the setup() function are performed in order. After the call to the size() function in the first statement, the second statement calls the balloon() function. At this point, the flow of execution jumps to the balloon() function, and a *copy* of each value in the argument list is stored in the corresponding parameter variable:

```
void balloon(int column, int row, int diameter)
{
   int lineLength = 50;

   line(column, row, column, row + lineLength);
   fill(255, 0, 0);
   ellipse(column, row, diameter, diameter);
}
```

The statements in the body of the balloon() function are now performed. As a result, a balloon is drawn at pixel column 75 and pixel row 50 with a diameter of 30 pixels:

After the last statement in the body of the `balloon()` function is performed, the flow of execution returns to the `setup()` function at the exact point at which the `balloon()` function was called:

```
void setup()
{
    size(200, 200);

    balloon(75, 100, 30);

}

void balloon(int column, int row, int diameter)
{
    int lineLength = 50;

    line(column, row, column, row + lineLength);
    fill(255, 0, 0);
    ellipse(column, row, diameter, diameter);
}
```

Flow of execution returns to the point from which the function was called.

Because all the statements in the `setup()` function have been performed, the execution of this program is complete.

As we have seen, we are allowed to declare *multiple* parameter variables in a function definition. This enables us to pass multiple items of information to such a function as arguments when we call the function.

Save the `BalloonDraw` program. We will return to it in a later section.

Other Special `void` Functions in Processing

As we have learned, the `setup()` function is a special `void` function because it is called *automatically* by Processing. There are a number of other special `void` functions in Processing that are called automatically or whenever a certain mouse or keyboard event takes place. These functions are fun to use for creating animation and interactivity in

Processing. However, because the somewhat "magical" behavior of these special functions is so different from the way most functions operate and is so specific to Processing, it is also the case that these special functions can be confusing to beginners who are trying to acquire an understanding of functions in general. This is the reason why they haven't been introduced in this book before now.

At the same time, these functions are so central to programming in Processing that if you hope to do more advanced programming in Processing, an introduction to these special functions is absolutely essential. Thus, the remainder of this chapter will give you an introduction to a few of these special void functions, particularly the fun and powerful draw() function. If you find these functions somewhat confusing, don't worry: they won't be used often in this book. However, if you do wish to learn more about these special void functions, you will find it easier to do so, now that you have learned about void functions in general.

A Special void Function: The draw() Function

As we have learned, the setup() function is a special Processing function. If we define the setup() function, Processing will *automatically* call the setup() function, just *once*.

The **draw()** function is another special Processing function. We have learned about *repetition* using the while and for statements. However, the draw() function offers a different kind of repetition. If we define the draw() function, then when we run our program and the defined actions of the setup() function are finished, Processing *automatically* calls the draw() function *repeatedly*, over and over again, until we stop our program. This kind of automatic repetition of the draw() function is not found in most programming languages, but it is central to creating animations and interactivity in Processing. As the name of the draw() function indicates, this is the location where we will typically put statements related to drawing on the canvas when we are creating an animation. However, in more general terms, the draw() function is where we will want to put any sequence of statements that we wish to have performed *repeatedly*.

Select **File > New** to start a new program. Save this program as **DrawDemo**.

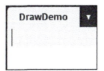

Let's start by defining the setup() function that simply displays a message to the console announcing that the setup() function was automatically called by Processing.

```
void setup()
{
  println("setup() was called");
}
```

When we run this program, Processing automatically calls the setup() function, and the following console output is produced:

setup() was called

Now, let's add the draw() function to our program. Like the setup () function, it is a void function with no parameters, so its *header* is

```
void draw()
```

And, like any function, we define the *body* of the draw() function between a pair of curly braces:

```
void draw()
{

}
```

As with any other function, we need to define the draw() function by inserting into its *body* the statements that we would like to have it perform. For example, let's have the draw() function simply generate a console message that is similar to that of our setup() function.

```
void draw()
{
  println("draw() was called");
}
```

Our program is now

```
void setup()
{
  println("setup() was called");
}

void draw()
{
  println("draw() was called");
}
```

Run this program, and as soon as you see console output, *immediately stop the program*. After stopping the program, *scroll* through the console output back to the beginning. You will see from the console output that Processing first called the setup() function, which produced the *first* line of output:

```
setup() was called
```

However, once the performance of the setup() function was finished, Processing then called the draw() function, which produced the *second* line of console output:

```
setup() was called
draw() was called ⬅
```

Then, once the performance of the draw() function was finished, Processing called the draw() function *again*, which produced the *third* line of output:

```
setup() was called
draw() was called
draw() was called ⬅
```

Then, once that performance of the draw() function was finished, Processing called the draw() function a *third* time, which produced the *fourth* line of output:

```
setup() was called
draw() was called
draw() was called
draw() was called ⬅
```

As you can see from the console output, Processing continued calling the draw() function again and again until you stopped the program.

```
setup() was called
draw() was called
draw() was called
draw() was called
        ⋮
draw() was called
```

Thus, when deciding how to use the draw() function, we need to consider two key facts:

1) The draw() function is called *automatically* by Processing, immediately *after* the call to the setup() function has finished.

2) The draw() function is called *repeatedly*, so the sequence of statements it contains is performed over and over again.

Save and close the DrawDemo program.

The `draw()` Function: A Graphical Example

Let's try using the `draw()` function for some repeated drawing. Select **File > New** to start a new program. Save this program as `Circles`.

We'll have the `setup()` function set the size of the canvas to 150 pixcls by 200 pixels:

```
void setup()
{
  size(150, 200);
}
```

Let's also have the `setup()` function draw a circle with a diameter of 30 pixels at a random location on the canvas:

```
void setup()
{
  size(150, 200);
  ellipse(random(width), random(height), 30, 30);
}
```

Run this program several times. As you can see, each time that we run this program, a circle is indeed drawn in a random location on the canvas:

Now, let's add the `draw()` function to our program.

```
void setup()
{
  size(150, 200);
  ellipse( random(width), random(height), 30, 30);
}

void draw()
{

}
```

As we have learned, when we run our program, Processing automatically calls the setup() function, once. When all the statements in the setup() function have been performed, Processing then calls the draw() function. When all the statements in the draw() function have been executed, Processing calls the draw() function *again*. This repeated and automatic calling of the draw() function continues until we stop our program.

Currently, our draw() function doesn't contain any statements. Thus, let's move the call to the ellipse() function from our setup() function to our draw() function:

```
void setup()
{
  size(150, 200);
}

void draw()
{
  ellipse(random(width), random(height), 30, 30);
}
```

When we run our program again, Processing now repeatedly draws a circle in a random location, over and over again, until we stop our program:

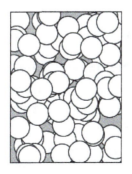

When we use the draw() function, we can also create programs that respond to the movement of the mouse pointer over the canvas. Processing provides two built-in variables of type int that store the current canvas pixel location of the *tip* of the mouse pointer each time the draw() function is called:

mouseX Stores the current canvas pixel *column* location of the tip of the mouse pointer

mouseY Stores the current canvas pixel *row* location of the tip of the mouse pointer

Notice that the *last letters* in these variable names are *capitalized*: mouse**X** and mouse**Y**. Let's try using the mouseX and mouseY variables. For example, we can change the call to the ellipse() function in our program so that it instead uses these two variables to specify the location of the circle that is drawn each time the draw() function is called:

```
void setup()
{
  size(150, 200);
}

void draw()
{
  ellipse(mouseX, mouseY, 30, 30);
}
```

When we run our program, circles are now drawn wherever we move the mouse pointer over the canvas:

Save the Circles program. We will return to it later in this chapter.

Animating the Balloon with the draw() Function

Let's try using the draw() function and the mouseX and mouseY variables in our balloon-drawing program. Return to the BalloonDraw program. Select **File > Save As** and resave this program as BalloonMove.

```
BalloonMove  ▾
void setup()
{
  size(200,
```

Because Processing's built-in mouseX and mouseY variables are both of type int, they may be used as *arguments* in a call to our balloon() function. For example, we can write

```
balloon(mouseX, mouseY, 30);
```

Let's insert a draw() function into our program, and let's place this call to the balloon() function inside the body of the draw() function:

```
void setup()
{
  size(200, 200);
}

void draw()
{
  balloon(mouseX, mouseY, 30);
}

void balloon(int column, int row, int diameter)
{
  int lineLength = 50;

  line(column, row, column, row + lineLength);
  fill(255, 0, 0);
  ellipse(column, row, diameter, diameter);
}
```

Now, when we run this program, each time the draw() function is performed, it calls the balloon() function. Thus, with each repetition of the draw() function, a balloon is drawn on the canvas at the current position of the mouse pointer:

Next, let's insert a call to the background() function inside the draw() function:

```
void draw()
{
  background(200, 200, 200);
  balloon(mouseX, mouseY, 30);
}
```

Now, with each repetition of the draw() function, the background() function is called first, generating a blank gray canvas. Then, the balloon() function is called, and a balloon is once again drawn at the current position of the mouse pointer. This creates the illusion that we are moving a *single* balloon around the canvas with our mouse.

And, if we make the following changes to our program,

```
void setup()
{
  size(200, 200);
  cursor(HAND);
}

void draw()
{
  background(200, 200, 200);
  balloon(mouseX, mouseY, 30);
}

void balloon(int column, int row, int diameter)
{
  int lineLength = 50;

  line(column, row, column, row - lineLength);
  fill(255, 0, 0);
  ellipse(column, row - lineLength, diameter, diameter);
}
```

then it seems as if a hand is holding the balloon at the end of the string:

It is sometimes hard to predict in advance the effect that the repeating draw() function will have, so programming in Processing using the draw() function often requires some trial and error. However, such unpredictability can also contribute to the fun that can result from using the draw() function!

Save and close the BalloonMove program.

Special void Functions in Processing Related to the Mouse and Keyboard

We have learned about two of Processing's special void functions: setup() and draw(). If we define a setup() function in a program, this function is *automatically* called *once* by Processing when we start the program. If we also define a draw() function, then this function is *automatically* called *repeatedly* by Processing after the setup() function is performed.*

If we define the draw() function, then we also have the option of defining any of several other special void() functions in Processing that are related to using the **mouse** and **keyboard**. The repetition of the draw() function (even if there are no statements in its body) keeps Processing "listening" for the occurrence of mouse and keyboard **events** such as moving the mouse, pressing a mouse button, or pressing a key on the keyboard.

For example, if we define the void **mousePressed()** function, then, each time that we press the mouse button while the mouse pointer is over the canvas, this function will be called *once* by Processing, and any statements in the body of this function will be performed. Afterward, Processing resumes its repeated calling of the draw() function. To illustrate, let's return to the Circles program and insert the following definition of a mousePressed() function that includes a call to the background() function:

---

* It is also permissible to define the draw() function *without* defining the setup() function.

```
void setup()
{
  size(150, 200);
}

void draw()
{
  ellipse(mouseX, mouseY, 30, 30);
}

void mousePressed()
{
  background(200, 200, 200);
}
```

Now, when we run this program, circles are drawn wherever we move the mouse pointer over the canvas, as before:

However, now, whenever we press the mouse button, the mousePressed() function is called. At this point, the flow of execution jumps to the mousePressed() function, and its call to the background() function is performed. Thus, all the pixels on the canvas are reset to the color specified in the call to the background() function. Afterward, the performance of the mousePressed() function is complete, so the flow of execution returns to the draw() function. Processing then resumes its repeated calling of the draw() function, so the repeated drawing of a circle at the current position of the mouse pointer starts all over again:

Processing also includes a similar void function named **keyPressed()**. As the name suggests, any time a key is pressed *while the canvas window is selected as the active window*, Processing will call the keyPressed() function a single time. At this point, the flow of execution jumps to the keyPressed() function, and the statements in its body are performed. The flow of execution then returns to the draw() function. Processing then resumes its repeated calling of the draw() function.

For example, if we change our program to use the keyPressed() function, then the blanking of the canvas using the background() function will instead take place when we press a key:

```
void setup()
{
  size(150, 200);
}

void draw()
{
  ellipse(mouseX, mouseY, 30, 30);
}

void keyPressed()
{
  background(200, 200, 200);
}
```

To learn more about other options related to the mouse and keyboard, see the **Mouse** and **Keyboard** entries in the **Input** section of Processing's Reference.

Save and close the Circles program.

Summary

In this chapter, we learned about creating our own void functions, which do *not* return a value.

- To create our own void functions in Processing, we need to be in *active mode*.

- Active mode includes the special void function setup(), which is *automatically* called *once* by Processing as soon as we start running our program.

- A function is created by defining its *header* and its *body*.

- A void function that we create is called in the same manner in which we call Processing's built-in void functions.

- A function can be *reused* by calling the function multiple times.

- We can define *parameter* variables for a function so that specific values can be passed to that function as *arguments*.

- Processing's special draw() function is *automatically* called *repeatedly* by Processing, as soon as the performance of the setup() function is completed.

- The mouseX and mouseY variables keep track of the current position of the mouse pointer over the canvas when the draw() function is used.

- Processing provides other special functions that can be used if the draw() function is defined: these include the mousePressed() and keyPressed() functions that are automatically called by Processing when the corresponding mouse or keyboard *event* takes place.

Exercises

1) Type in the following program:

```
void setup ()
{

}

void smiley ()
{
  println( ":-)" );
}
```

Then, modify this program so that the setup() function calls the smiley() function. When you run this program, you should see the following output:

:-)

2) Modify the previous program so that multiple calls to the smiley() function produce the following output:

:-)

:-)

:-)

:-)

:-)

3) Modify the previous program to instead use a for statement containing a call to the smiley() function that produces the same output.

4) Write a `void` function named `qwerty()` that displays the letters on each of the three lines of a QWERTY keyboard. Have the `setup()` function display "Keyboard Letters" to the console and then call the `qwerty()` function. Running the program should produce the following console output:

```
Keyboard Letters:
QWERTYUIOP
ASDFGHJKL
ZXCVBNM
```

5) Write a `void` function named `days()` that displays the days of the week in the console. Have the `setup()` function display "Days of the week:" in the console and then call the `days()` function. Running the program should produce the following console output:

```
Days of the week:
Sunday
Monday
Tuesday
Wednesday
Thursday
Friday
Saturday
```

6) In Processing, the `month()` function returns the month number. The `day()` function returns the day of the month. The `year()` function returns the year. For example, the following statement would display the current month to the console:

```
println("Month: " + month() );
```

Write a `void` function named `today()` that displays the current month, day, and year to the console. Have the `setup()` function display "TODAY:" to the console and then call the `today()` function. Running the program should produce console output like the following:

```
TODAY:
Month: 5
Day: 28
Year: 2017
```

7) Write a void function named `hashtags ()` that displays a hashtag on a single line in the console but does not advance to a new console line afterward. Have the `setup()` function display "hashtags:" to the console and then call the `hashtags ()` function. Running the program should produce the following console output:

```
hashtags:
#
```

8) Modify the previous program so that the `hashtags()` function has a single int parameter variable named `quantity` that specifies the number of hashtags that the function should display. Insert a `for` statement into the `hashtags()` function to accomplish this. Running the program with `hashtags(5)` as the call to the function should produce the following output:

```
hashtags:
#####
```

Running the program with `hashtags(10)` as the call to the function should produce the following output:

```
hashtags:
##########
```

9) Write a void function named `square ()` that draws a square with a side length of 50 pixels in the upper-left corner of a 200-pixel by 200-pixel canvas. Call this function from the `setup()` function. Running this program should produce the following output:

10) Modify the program in the preceding exercise so that the square() function has an int parameter variable side to specify the side length of the square. Running the program with square(100) as the call to the function should produce the following output:

11) Modify the program in the preceding exercise so that the square() function also has an int parameter variable column and an int parameter variable row for specifying the column and row of the upper-left corner of the square. Running the program with square(100, 25, 50) as the call to the function should produce the following output:

12) Modify the previous program by inserting a `for` statement into the `setup()` function that calls the `square()` function five times, using 20 as the side length and random values for the column and row numbers. Running the program should produce output like the following:

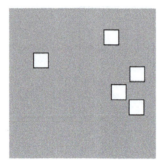

13) Write a `void` function named `yoyo()` that draws a line from the top center of a 200-pixel by 300-pixel canvas to the center of the canvas and also draws a circle with a diameter of 50 pixels in the center of the canvas. Call this function from the `setup()` function. Running this program should produce output like the following:

14) Modify the previous program so that the yoyo () function has an int parameter variable named row that specifies the vertical position of the yoyo. Running the program with yoyo (220) as the call to the function should produce the following output:

15) Modify the previous program by doing the following:

a) Declare an `int` variable named `randomRow` in the `setup()` function.

b) Assign `randomRow` a random integer from 0 to the height of the canvas.

c) Use `randomRow` as the argument in the call to the `yoyo()` function.
Running this program should place the yoyo at a different vertical position each time.
Sample runs

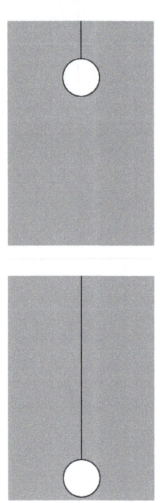

CHAPTER 8

Creating Functions
That Return a Value

In this chapter, we will learn how to create our own functions that return a value.

(Non-void) Functions

In Chapter 7, we considered a `void` function as a *sequence* of individual actions that have been given a name and that, together, comprise some desirable *overall* action. As an analogy, we considered how a parent's general instruction to a teen to "get ready for school" is actually a concise way of referring to a whole sequence of activities that need to be performed.

Now, let's consider a slightly different analogy. Suppose that, each year, a parent instructs a teen:

"Find out how much it will cost for you to play soccer this season."

Both the parent and teen know that this general instruction actually translates into a whole sequence of individual instructions, such as

- "Find out the cost of the cleats you need."
- "Find out the cost of the shin guards you need."
- "Find out the cost of the soccer ball you need."
- "Find out the cost of the team uniform."
- "Add up these costs."
- "Report back the total cost."

Because the individual activities of this routine are predefined, the parent can ask the teen to perform this entire sequence simply by saying, "Find out how much it will cost for you to play soccer this season." However, notice that, unlike the "Get ready for school" analogy, the sequence of activities in this second analogy ends by requiring the teen to *bring back* an item of information—namely, the total cost of playing soccer this season.

In Processing, a *non-*void **function** is similar to this second analogy in that a named sequence of statements is performed, but a resulting item of information described by the general instruction is also *brought back*. Such non-void functions that *return a value* are more commonly known simply as "functions."

In this book, we have used a number of built-in Processing functions that return a value. For example, behind the scenes, the sqrt() function is defined as a sequence of steps that comprise the overall action of calculating the square root of whatever value we supply as an argument to this function. This square root result is then *returned* to us as a float value.

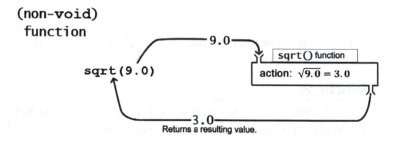

Remember also that, when we call a non-void function, we need to make immediate *use of the returned value*. For example, we know that if we write the statement

```
sqrt(9.0);
```

then the square root returned by this function call is *lost*. We have learned that a good technique for making immediate use of the value returned by a function is to assign this value to a *variable*. For example, we can store the square root returned by the sqrt() function in a float variable, root:

```
float root;
root = sqrt(9.0);
```

This way, the resulting value returned by the function is stored in a variable and is available to us for further use. For instance, we could use this variable to display the returned value in the console:

```
float root;
root = sqrt(9.0);
println("Square root: " + root);
```

Similarly, as we will see, when we create *our own* function that returns a value, we will need to make immediate use of the value returned whenever we call such a function.

Creating Our Own Function That Returns a Value

In Processing, we can create our own non-`void` functions that return a value. Doing so is very similar to creating `void` functions, but there are also some key differences.

For example, suppose we once again write a program to simulate rolling a die. Select **File > New** to create a new program. Save this program as **RollingFunction**.

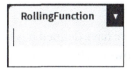

As when we create `void` functions, we need to work in Processing's *active mode* in order to create functions that return a value. Thus, let's begin a definition of the `setup()` function by writing the following:

```
void setup()
{

}
```

As before, to simulate rolling a die, we will generate a random integer from 1 to 6, store this result in a variable, and then display the value of this variable in the console. To achieve this, we'll add the following statements to the body of our `setup()` function:

```
void setup()
{
  int die;

  die = int( random(1, 7) );

  println(die);
}
```

When we run this program, we see an integer from 1 to 6 displayed in the console:

```
4
```

However, suppose we want our program to simulate the roll in a game that uses *five* dice. One way to achieve this would be by using five similar statements in our program:

```
die1 = int( random(1, 7) );
die2 = int( random(1, 7) );
die3 = int( random(1, 7) );
die4 = int( random(1, 7) );
die5 = int( random(1, 7) );
```

However, it's somewhat cumbersome to write nested function calls to the `int()` and `random()` functions each time we want to simulate rolling a die. Let's instead create a *function* that will perform this simulation.

As we have learned, when defining a function, we start with our function *header*, which requires *three* items of information:

1) The *type of value to return*.

   Thus far, each function that we have defined has been a `void` function, which means it does not return a value. However, for a function that returns a value, we do need to define the *type* of value *returned*. In our function that simulates rolling a die, we need to have a value returned. Specifically, we need the function to return the integer from 1 to 6 that is chosen at random, so the type of value we wish to return is `int`. Thus, we begin our function header with

   ```
   int
   ```

2) The function's *name*.

   Let's use "`roll`" as the name of this function. Thus, our header is now

   ```
   int roll
   ```

3) A list of the function's *parameters*.

   These are the items of information that the function requires in order to perform the statements contained in its body. Our die-simulating function does not currently require any such items, so we simply write an empty pair of parentheses:

   ```
   int roll()
   ```

With the function header complete, we'll next define the *body* of our `roll()` function, starting with a pair of curly braces:

```
int roll()
{

}
```

Our program is now

```
void setup()
{
  int die;

  die = int( random(1, 7) );

  println(die);
}
```

```
int roll()
{

}
```

Next, let's insert into the body of our `roll()` function that we will need to simulate a roll of a die. We will adopt the practice of declaring a *variable* to store the resulting value that will be returned by a function. In this particular case, let's declare a variable named `result` to store the random integer value from 1 to 6 that will need to be returned as the result of each call to the `roll()` function:

```
int roll()
{
  int result;
}
```

As we know, a variable declared inside a function is a *local* variable and can only be used *inside* that function. Here, `result` is a local variable, and its scope (range of use) is limited to the `roll()` function. The next step in defining the body of our `roll()` function is selecting a random `int` value from 1 to 6 and assigning it to the `result` variable. To achieve this, we add the following statement:

```
int roll()
{
  int result;
  result = int( random(1, 7) );
}
```

You may have noticed that *underlining* has appeared in the header of our function definition:

```
int roll()
{
  int result;
  result = int( random(1, 7) );
}
```

If we click on this underlining, we see that the corresponding error message in the Message Area is

This method must return a result of type int.

We are seeing this error message because our `roll()` function is *not* a void function; rather, its return type is `int`. Whenever we specify a return type in the function header other than void, we have made a commitment to *returning* a value of that specified *type*. Thus, our roll() function *must return* a value of type int.

We return a value using a **return statement**. The basic form of a return statement consists of the keyword return, followed by the resulting value that is to be returned:

```
return value-to-return;
```

For our roll() function, we need to return the randomly chosen int value from 1 to 6 that we have stored in the result variable. Thus, we will insert the following statement at the *end* of the body of our function:

```
int roll()
{
  int result;
  result = int( random(1, 7) );
  return result;
}
```

Adding this return statement now ensures that a copy of the value stored in the result variable will be returned whenever the roll() function is called. Thus, after we make these changes, both the underlining and error message disappear. A return statement also causes the flow of execution to return *immediately* to the point from which the function was called. For this reason, the return statement is usually the *last* statement in the body of a function definition.

Our program is currently

```
void setup()
{
  int die;

  die = int( random(1, 7) );

  println(die);
}
int roll()
{
  int result;
  result = int( random(1, 7) );
  return result;
}
```

Our roll() function is ready to use. How should we call our new roll() function from the setup() function? Remember, our roll() function *returns a value*. Thus, any statement that contains a call to this function needs to make *immediate use* of this returned value. As usual, we will do this by assigning the returned value to a *variable*. In this particular case, we will assign the returned result to the die variable, replacing the nested calls to the int() and random() functions:

```
void setup()
{
  int die;

  die = roll();

  println(die);
}
```

As we have learned, storing this returned result in the die variable enables us to make further use of the returned value—for example, in the subsequent call to the println() function:

```
void setup()
{
  int die;

  die = roll();

  println(die);
}
```

Our full program is now

```
void setup()
{
  int die;

  die = roll();

  println(die);
}
int roll()
{
  int result;
  result = int( random(1, 7) );
  return result;
}
```

When we run this program, we see that a value from 1 to 6 is output to the console:

5

Let's carefully trace the flow of execution in our current program. When we run our program, the setup() function is called automatically, and the statements in the body of this function are performed. The *first* statement of our setup() function declares our die variable:

```
void setup()
{
  int die;

  die = roll();

  println(die);
}
```

int [_____]
die

The *second* statement in the setup() function is an assignment statement:

```
void setup()
{
  int die;

  die = roll();

  println(die);
}
```

Remember, in an assignment statement, the right-hand side of the assignment operator (=) is always evaluated first.

$$\textbf{die} = \boxed{\textbf{roll()}};$$

In this particular assignment statement, the right-hand side consists of a call to the roll() function. Thus, in order to evaluate the right-hand side, the flow of execution jumps to the roll() function.

```
void setup()
{
  int die;

  die = roll();

  println(die);
}

int roll()
{
  int result;
  result = int( random(1, 7));
  return result;
}
```

The statements in the body of the roll() function are now performed in sequential order. In the *first* statement in the body of this function, the int variable result is declared:

```
int roll()
{
  int result;
  result = int( random(1, 7) );
  return result;
}
```

In the *second* statement in the body of the roll() function, the result variable is assigned an int value from 1 to 6:

```
int roll()
{
  int result;
  result = int( random(1, 7) );
  return result;
}
```

The *third* and final statement of the roll() function is a return statement:

```
int roll()
{
  int result;
  result = int( random(1, 7) );
  return result;
}
```

This return statement causes the flow of execution to return immediately to the point in the setup() function from which the roll() function was called, bringing along a copy of the value that was stored in the result variable (e.g., 5).

```
void setup()
{
  int die;

  die = roll();               5

  println(die);
}

int roll()
{
  int result;
  result = int( random(1, 7) );
  return result;
}
```

With this returned value, the evaluation of the right-hand side is now complete. We can visualize the result of this evaluation as

die = 5;

With the evaluation of the right-hand side complete, the assignment operation can now be performed. We can visualize the result of this assignment operation as

die = 5;   int | 5 |
                    die

Next, the performance of the statements in the setup() function now continues with the next statement:

```
void setup()
{
  int die;

  die = roll();

  println(die);
}
```

As a result of this statement, the value stored in the die variable is output to the console:

```
5
```

The end of the body of the setup() function has been reached. Thus, the execution of this program is now finished.

Save the RollingFunction program. We will continue working with it in the next section.

Multiple Function Calls

Return to the RollingFunction program. Select **File** > **Save As** and resave this program as **MoreRolling**.

Suppose that we would like to simulate the action of rolling *two* dice. For this, we can use two variables in our setup() function to store the value of each die:

```
int die1;
int die2;
```

We can then use our roll() function to assign a value to each of these two variables:

```
die1 = roll();
die2 = roll();
```

Afterward, we can display the values of both of these variables in the console by making the following changes to our call to the println() function:

```
println(die1 + " " + die2);
```

After we make these changes to our setup() function, our program consists of the following:

```
void setup()
{
  int die1;
  int die2;

  die1 = roll();
  die2 = roll();

  println(die1 + " " + die2);
}

int roll()
{
  int result;
  result = int( random(1, 7) );
  return result;
}
```

Now, each time the program is run, the roll() function is called *twice*, once for each die, and the values of the two dice are displayed in the console:

```
2 6
```

Suppose that we would like to create a game in which we roll these two dice *repeatedly* until the dice *match*. In other words, we would like to roll *as long as* the two dice *do not* match. To achieve this repetition, we can insert the following do-while statement:

```
void setup()
{
  int die1;
  int die2;
```

```
do
{
  die1 = roll();
  die2 = roll();

  println(die1 + " " + die2);
}
while (die1 != die2);

}

int roll()
{
  int result;
  result = int( random(1, 7) );
  return result;
}
```

Now, when we run this program, *both* calls to the roll() function are repeated until the two values returned match each other. As a result, we see console output like the following:

```
1 3
6 2
2 6
3 5
2 1
4 4
```

Save and close this MoreRolling program.

Adding a Parameter to a Function That Returns a Value

We learned how to define parameter variables for void functions in Chapter 7. We can follow the same steps to define parameter variables for functions that return a.

For example, let's return to our RollingFunction program that simulates the roll of a single die:

```
void setup()
{
  int die;

  die = roll();

  println(die);
}
```

```
int roll()
{
  int result;
  result = int( random(1, 7) );
  return result;
}
```

Select **File** > **Save As** and resave this program as **ParameterRolling**.

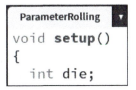

Games such as Monopoly involve rolling *multiple* dice and using only the *total* of the dice values. Let's modify our `roll()` function so that it can simulate such a roll of multiple dice without requiring us to make multiple calls to this function. For this, we'll need to add a *parameter* variable to our `roll()` function so that the number of dice can be passed as an *argument* when this function is called:

```
int roll(int numberOfDice)
{
  int result;
  result = int( random(1, 7) );
  return result;
}
```

To enable our `roll()` function to simulate the roll of a specified number of dice, we will instead have the function simulate *repeatedly* rolling a *single* die that number of times and calculate the *total* of these multiple rolls. For example, to simulate rolling three different dice, our `roll()` function will instead simulate rolling a single die three times and then return the total of these three rolls.

To use our `result` variable to calculate a running total of the repeated rolls using, we'll need to start by initializing this variable to zero each time the `roll()` function is called. We can do this in a separate statement:

```
int roll(int numberOfDice)
{
  int result;
  result = 0;
  result = int( random(1, 7) );
  return result;
}
```

However, to keep our code more concise, let's declare and initialize the `result` variable in a single statement:

```
int roll(int numberOfDice)
{
  int result = 0;
  result = int( random(1, 7) );
  return result;
}
```

Next, we'll insert a `for` loop so that our simulation of rolling a single die is repeated the number of times specified by the `numberOfDice` parameter variable:

```
int roll(int numberOfDice)
{
  int result = 0;

  for (int count = 1; count <= numberOfDice; count = count + 1)
  {
    result = int( random(1, 7) );
  }

  return result;
}
```

In previous examples that have calculated a running total, we have declared a variable specifically for this purpose. Such a totaling variable must be declared and initialized before the loop of values during the repetitions of a loop. Each time the body of the loop is performed, we add a new amount to this totaling variable and then assign this sum as the new value of the totaling variable. To achieve this, we insert into the body of our loop a statement that has the following form:

```
total = total + amount;
```

By inserting a statement of this form, we ensure that, when the loop stops repeating, the totaling variable will indeed contain the correct total.

In this particular case, we will use the `result` variable to do the totaling. The amount we will be adding to the running total each time the `for` loop repeats is the random die value that is generated. Thus, the statement we need to keep a running total of the repeated die rolls is

```
result = result + int( random(1, 7) );
```

Accordingly, we will simply change the body of our for statement to the following:

```
int roll(int numberOfDice)
{
  int result = 0;

  for (int count = 1; count <= numberOfDice; count = count + 1)
  {
    result = result + int( random(1, 7) );
  }
  return result;
}
```

Now, each time the for loop is performed, the value of result will be increased by the value of the current die roll. Thus, when the for loop is done repeating, the result variable will indeed contain the total of all the die rolls.

The new version of our roll() function is now ready to be used. However, we need to modify our setup() function so it is compatible with the changes that we've made in our roll() function.

First, let's change the name of the die variable to dice, which is more fitting for simulating the rolling of multiple dice:

```
void setup()
{
  int dice;

  dice = roll();

  println(dice);
}
```

We also see in the setup() function that the call to the roll() function is *underlined*.

```
void setup()
{
  int dice;

  dice = roll();

  println(dice);
}
```

When we click on this underlining, we see that the corresponding error message is

The function roll() expects parameters like: "roll(int)"

As this error message indicates, the header of the modified version of our `roll()` function definition now contains the `numberOfDice` *parameter* variable, which is of type `int`:

```
int roll(int numberOfDice)
```

Thus, whenever we call the new version of our `roll()` function, we now need to supply a corresponding *argument* of type `int`—namely, the number of dice to be rolled. For example, to simulate the rolling of two dice, we can change this function call to

```
dice = roll(2);
```

After making these changes to the `setup()` function, our program is now

```
void setup()
{
  int dice;

  dice = roll(2);

  println(dice);
}

int roll(int numberOfDice)
{
  int result = 0;

  for (int count = 1; count <= numberOfDice; count = count + 1)
  {
    result = result + int( random(1, 7) );
  }

  return result;
}
```

When we run this new version of our program, the range of the randomly chosen value displayed in the console now depends on the number of dice specified. For example, we are currently simulating the roll of two dice. Thus, the value chosen is in the range from 2 to 12:

```
11
```

Let's trace the flow of execution for this program. When we start the program, the `setup()` function is called automatically, and the statements in its body are performed. The first statement in the `setup()` function declares the `dice` variable. We can visualize this variable as

int

dice

The second statement in the body of the setup() function is an assignment statement.

```
void setup()
{
  int dice;

  dice = roll(2);

  println(dice);
}
```

As always, the right-hand side of the assignment operator (=) is evaluated first.

$$\texttt{dice} = \boxed{\texttt{roll(2)}};$$

In this particular case, the right-hand side contains a call to the roll() function. Thus, at this point, the flow of execution jumps to the roll() function, bringing along a copy of the int value that we supplied as an argument to specify the number of dice—in this particular case, 2. We can visualize this as

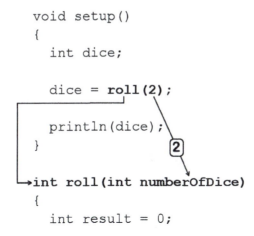

The argument that we supplied in the function call is stored in the roll() function's parameter variable, numberOfDice. We can visualize this as

```
        ②
int roll(int numberOfDice)
{
  int result = 0;

  for (int count = 1; count <= numberOfDice; count = count + 1)
  {
    result = result + int( random(1, 7) );
  }

  return result;
}
```

The statements in the body of the `roll()` function are now performed. The first statement declares and initializes the `result` variable. In the second statement, the body of the `for` loop is performed the same number of times as the argument value stored in `numberOfDice`.

```
for (int count = 1; count <= numberOfDice; count = count + 1)
{
  result = result + int ( random (1, 7) );
}
```

In this particular case, the value of `numberOfDice` is 2; thus, the body of the `for` statement is performed two times.

Each time the body of the `for` statement is performed, the simulated roll of a die is performed and added to the running total stored in the `result` variable.

```
for (int count = 1; count <= numberOfDice; count = count + 1)
{
  result = result + int ( random (1, 7) );
}
```

When the `for` statement is done repeating, `result` always contains a value ranging from the number of dice to the number of dice multiplied by 6. In this case, the number of dice is two, so the value of `result` will always be between 2 and 12 (for example, 11).

Next, the `return` statement is encountered:

```
return result;
```

When the `return` statement is performed, the flow of execution immediately returns from the `roll()` function to the point in the `setup()` function from which it was called, bringing along a copy of the `int` value that is stored in the `result` variable. We can visualize this ras

```
  dice = roll(2); ←——————11——————┐
                                   │
  println(dice);                   │
}                                  │
                                   │
int roll(int numberOfDice)         │
{                                  │
  int result = 0;                  │
                                   │
  for (int count = 1; count <= numberOfDice; count = count + 1)
  {                                │
    result = result + int ( random (1, 7) );
  }                                │
                                   │
  return result; ──────────────────┘
}
```

The evaluation of the right-hand side of the assignment statement is now complete. Thus, the assignment statement essentially becomes

$$\texttt{dice = }\boxed{11};$$

The assignment operation is now performed, storing the returned `int` value in the `dice` variable on the left-hand side. We can visualize this assignment operation and its result as

$$\texttt{dice = }\boxed{11}; \qquad \texttt{int}\ \boxed{\textbf{11}}$$
$$\texttt{dice}$$

The performance of the body of the `setup()` now resumes with the next statement:

```
void setup()
{
  int dice;

  dice = roll();
  println(dice);
}
```

Thus, the total value of the number of dice rolled is displayed to the console.

```
11
```

This statement was the last statement in the `setup()` function, so the execution of our program is complete.

Save and close the `ParameterRolling` program.

Revisiting the Right Triangle Example

When working with `void` functions, we saw that we can define functions that have *multiple* parameters. This is also an option for functions that return a value.

To illustrate, let's return to the version of our `Triangle` program that we last used in Chapter 3. Select **File** > **Save As** and resave this program as **HypFunction**.

We need to convert the current version of our program, which was written in *static* mode, to a program in *active* mode. We can do this by simply placing the current statements of our program inside the body of the `setup()` function. Thus, our program is now

```
void setup()
{
  size(300, 300);

  int leg1;
  leg1 = 150;

  int leg2;
  leg2 = 200;

  triangle(0, 0, leg1, 0, 0, leg2);

  float hypotenuse;
  hypotenuse = sqrt((leg1 * leg1) + (leg2 * leg2));
  println("Hypotenuse: " + hypotenuse);
}
```

As before, when we run this program, we see a right triangle with the specified leg lengths drawn on the canvas:

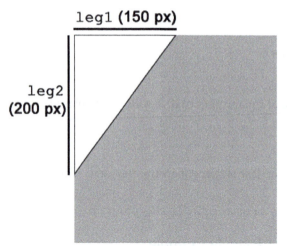

We also see the length of the hypotenuse displayed in the console:

Hypotenuse: 250.0

Let's create a function that, given the two leg lengths, calculates and returns the hypotenuse. As usual, we start the *header* of a function definition by specifying the *three* required items of information.

1) The *type of value returned*.

   As we have seen when we run this program, the calculated hypotenuse usually has nonzero decimal places, so we should have this function return a value of type float. Thus, we begin our function header with

   ```
   float
   ```

2) The function *name*.

Hypotenuse is commonly abbreviated *hyp*, so let's use this as the name of our function. Thus, we add this name to the function header:

```
float hyp
```

3) A list of the function's *parameters*.

This function will need us to send it two *argument* values, one for each leg of the triangle. Thus, we need to add to our header the two *parameter* variable declarations that will store these two arguments. To avoid confusion, let's use different variable names from those that we are using in the setup() function to store the two triangle legs.* Thus, we'll write

```
float hyp(int legA, int legB)
```

With the header of the hyp() function complete, we are now ready to begin our definition of the *body* of this function:

```
float hyp(int legA, int legB)
{

}
```

As usual, we will declare a variable named result in which to store the resulting value that will be returned by the function. In the case of this hyp() function, the resulting hypotenuse value that we will be returning will be of type float, so we'll write the following variable declaration:

```
float hyp(int legA, int legB)
{
    float result;
}
```

We can adapt the formula that we used in the setup() function to calculate the hypotenuse using our two parameter variables and assign the result to our result variable:

---

* Actually, we could have used the *same* variable names, because now that we are working in *active mode*, leg1 and leg2 are now *local* variables: their *scope* (i.e., the portion of the program where they can be used) is limited to the setup() function. This means that we could have declared local variables within the hyp() function that had the same names, because the scope of these variables would be limited to the hyp() function. Because each of these pairs of variables have a separate scope, reusing these variable names would not have caused an error. However, avoiding the reuse of different variable names will usually make your program a bit more readable.

```
float hyp(int legA, int legB)
{
  float result;
  result = sqrt((legA * legA) + (legB * legB));
}
```

We also need to return this calculated result:

```
float hyp(int legA, int legB)
{
  float result;
  result = sqrt((legA * legA) + (legB * legB));
  return result;
}
```

We can now use our new hyp() function in the setup() function simply by making the following change to the assignment statement:

```
hypotenuse = hyp(leg1, leg2);
```

Thus, our entire program is now

```
void setup()
{
  size(300, 300);

  int leg1;
  leg1 = 150;

  int leg2;
  leg2 = 200;

  triangle(0, 0, leg1, 0, 0, leg2);

  float hypotenuse;
  hypotenuse = hyp(leg1, leg2);
  println("Hypotenuse: " + hypotenuse);
}
float hyp(int legA, int legB)
{
  float result;
  result = sqrt((legA * legA) + (legB * legB));
  return result;
}
```

When we run this program, the same triangle is drawn as before. However, the int variables leg1 and leg2 are now supplied as *arguments* in the call to the hyp() function.

Thus, when the flow of execution jumps to the hyp() function, the int values of leg1 and leg2 become the values of the int *parameter* variables legA and legB, respectively. We can visualize this as

```
void setup()
{
  size(300, 300);

  int leg1;
  leg1 = 150;

  int leg2;
  leg2 = 200;

  triangle(0, 0, leg1, 0, 0, leg2);

  float hypotenuse;
  hypotenuse = hyp(leg1, leg2);
  println("Hypotenuse: " + hypotenuse);
}
                     150    200

float hyp(int legA, int legB)
{
  float result;
  result = sqrt((legA * legA) + (legB * legB));
  return result;
}
```

The hyp() function then uses the values of the legA and legB variables to calculate the length of the hypotenuse and stores this length in the result variable.

```
float hyp(int legA, int legB)
{
  float result;
  result = sqrt((legA * legA) + (legB * legB));
  return result;
}
```

Thus, the expression on the right-hand side of the assignment operator (=) essentially becomes

```
result = sqrt((150 * 150) + (200 * 200));
```

And, when this expression is evaluated, this assignment statement essentially becomes

```
result = 250.0;
```

After this assignment statement is performed, the `return` statement is encountered, and the value stored in `result` is returned. We can visualize this `return` as

```
void setup()
{
  size(300, 300);

  int leg1;
  leg1 = 150;

  int leg2;
  leg2 = 200;

  triangle(0, 0, leg1, 0, 0, leg2);

  float hypotenuse;
  hypotenuse = hyp(leg1, leg2);
  println("Hypotenuse: " + hypotenuse);
}

float hyp(int legA, int legB)
{
  float result;
  result = sqrt((legA * legA) + (legB * legB));
  return result;
}
```

250.0

Thus, the assignment statement containing the call to the `hyp()` function now essentially becomes

```
hypotenuse = 250.0;
```

and the value of the `hypotenuse` variable becomes `250.0`, which we can visualize as

As a result, when the next statement is performed,

```
println("Hypotenuse: " + hypotenuse);
```

we see the same output of the hypotenuse to the console as before:

Hypotenuse: 250.0

The performance of the `setup()` function is now finished. Thus, the execution of our program is now complete.

Let's now modify this program so that it selects the length of the two legs at *random*. We can do this by making the following changes to the `setup()` function:

```
void setup()
{
  size(300, 300);

  int leg1;
  leg1 = int( random(width) );
  println("leg1: " + leg1);

  int leg2;
  leg2 = int( random(height) ) ;
  println("leg2: " + leg2);

  triangle(0, 0, leg1, 0, 0, leg2);

  float hypotenuse;
  hypotenuse = hyp(leg1, leg2);
  println("Hypotenuse: " + hypotenuse);
}
```

Now, when we run this program, the right triangle drawn on the canvas has legs of random lengths:

These randomly selected leg lengths are displayed in the console, along with the calculated length of the hypotenuse:

```
leg1: 294
leg2: 248
Hypotenuse: 384.6297
```

Save the `HypFunction` program. We will continue using it in the next section.

Multiple Functions in the Same Program

It is sometimes desirable to have *multiple* function definitions within the same program. For example, given a triangle's base length (*b*) and its altitude (*a*),

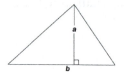

the formula for the area of the triangle is $A=\frac{1}{2}ba$. This is not a complicated formula. However, to see how a single program can contain *multiple* function definitions, let's add a function to our program that will calculate the area of a triangle.

Return to the `HypFunction` program. Select **File > Save** and resave this program as **TwoFunctions**.

We can define a *second* function in our current program simply by creating another function definition. In this particular case, we'll add a new definition of a function named `triangleArea()` after the current definition of our `hyp()` function.

As usual, we start the *header* of a function definition by specifying the *three* required items of information.

1) The *type of value returned*.

   Even if the base and altitude of a triangle are given as `int` values, the formula for the area of a triangle can produce a nonzero decimal place (`.5`) if the product of the base and altitude is an odd number. Thus, we should have this function return a value of type `float`. For this, we need to begin our function header with:

   float

2) The function *name*.

   Let's use `triangleArea()` as the name of our function:

   float **triangleArea**

3) The list of *parameters*.

   This function will need us to send it two *arguments*, the length of the base of the triangle and the altitude of the triangle. Thus, we need to add to our header the two *parameter* variable declarations that will store these two arguments:

```
float triangleArea(int base, int altitude)
```

With the header of the `triangleArea()` function complete, we are now ready to start the definition of the *body* of this function:

```
float triangleArea(int base, int altitude)
{

}
```

As usual, we will declare a variable named `result` in which to store the resulting value that will be returned by the function. In the case of the `triangleArea()` function, the resulting area value that we will return is of type `float`:

```
float triangleArea(int base, int altitude)
{
  float result;
}
```

We can now use our triangle area formula to assign a value to the `result` variable. We might be tempted to write the following statement:

```
float triangleArea(int base, int altitude)
{
  float result;
  result = 1 / 2 * base * altitude;
}
```

However, the operation

```
1 / 2
```

is *integer* division and evaluates to 0. Thus, if we use this statement, our triangle's area will always be zero. We might also be tempted to write the following statement:

```
float triangleArea(int base, int altitude)
{
  float result;
  result = base * altitude / 2;
}
```

However, the expression on the right-hand side of the assignment operator (=) in this statement,

```
base * altitude / 2;
```

consists entirely of int operands. Thus, the product of base and altitude is an int value. This means that when this product is divided by the int value 2, the operation is once again *integer* division. Thus, if the product of base and altitude is an *odd* number, an incorrect result will be produced as a result of integer division by 2. We need to make sure that integer division is not used in our formula. In this particular case, the simplest way to avoid this is to multiply by the float value 0.5 instead of dividing by the int value 2:

```
float triangleArea(int base, int altitude)
{
  float result;
  result = base * altitude * 0.5;
}
```

This statement will calculate the correct area, even when the product of base and altitude is an odd number. Now that we have a statement that correctly calculates the value to assign to result, we simply need to return this value:

```
float triangleArea(int base, int altitude)
{
  float result;
  result = base * altitude * 0.5;
  return result;
}
```

Our triangleArea() function is complete.*

We can now call our new triangleArea() function from the setup() function. In our function call, we need to supply two int arguments that will become the values of the base and altitude *parameter* variables that we have defined in the header of the triangleArea() function. In the case of a right triangle, the two legs can be used to specify the base and height of a triangle.

---

* In cases when the calculation of the value to return is a simple process that can be done in a single statement, we do have the option of calculating and returning this resulting value *without* using a variable such as result. For example, we could reduce the triangleArea() function to the following:

```
void triangleArea(int base, int altitude)
{
 return base * height * 0.5;
}
```

However, this form is best used when the formula for calculating the returned value is one that is simple enough to be *easily* understood by someone reading your code. Avoid creating lengthy, complex, hard-to-read return statements. Instead, use a local variable in the manner in which we have used the result variable whenever doing so makes your function definition easier to understand.

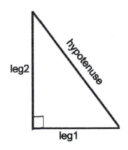

Thus, our function call can be the following:

```
triangleArea(leg1, leg2);
```

We'll add a declaration of a variable, area, to hold the triangle area that will be returned by our new triangleArea() function:

```
float area;
```

We can now write:

```
float area;
area = triangleArea(leg1, leg2);
```

Finally, let's display the triangle area in the console:

```
float area;
area = triangleArea(leg1, leg2);
println("Area: " + area);
```

Our entire program is now

```
void setup()
{
  size(300, 300);

  int leg1;
  leg1 = int( random(width) );
  println("leg1: " + leg1);

  int leg2;
  leg2 = int( random(height) ) ;
  println("leg2: " + leg2);

  triangle(0, 0, leg1, 0, 0, leg2);

  float hypotenuse;
  hypotenuse = hyp(leg1, leg2);
  println("Hypotenuse: " + hypotenuse);
```

```
  float area;
  area = triangleArea(leg1, leg2);
  println("Area: " + area);

}

float hyp(int legA, int legB)
{
  float result;
  result = sqrt((legA * legA) + (legB * legB));
  return result;
}

float triangleArea(int base, int altitude)
{
  float result;
  result = base * altitude * 0.5;
  return result;
}
```

We now have *two* function definitions within the same program. As a result, *both* of these functions are now available for us to use at *any* point in this program.

Having two function definitions in our program does make the flow of execution a bit more complicated when we run our program. Let's examine this a bit more closely. In our current program, we call each of our two defined functions *once* from the setup() function. As a result, the flow of execution jumps out of the setup() function *twice*. The *first* jump happens when the hyp() function is called. At this point, the flow of execution jumps to the hyp() function, and the values of the variables supplied as arguments in the function call are passed as the corresponding values of the hyp() function's parameter variables. We can visualize this as

```
  hypotenuse = hyp(leg1, leg2);
  println("Hypotenuse: " + hypotenuse);

  float area;
  area = triangleArea(leg1, leg2);
  println("Area: " + area);

                       263        187

}

float hyp(int legA, int legB)
{
  float result;
  result = sqrt((legA * legA) + (legB * legB));
  return result;
}
```

When the performance of the hyp() function is complete, the flow of execution returns to the exact point in the setup() program from which the hyp() function was called. The float value resulting from this function call is included in this return. We can visualize this return as

```
hypotenuse = hyp(leg1, leg2);
println("Hypotenuse: " + hypotenuse);

float area;
area = triangleArea(leg1, leg2);
println("Area: " + area);

}                                                322.7042


float hyp(int legA, int legB)
{
    float result;
    result = sqrt((legA * legA) + (legB * legB));
    return result;
}
```

The *second* jump happens when the statement containing the call to the triangleArea() function is performed. At this point, the flow of execution jumps to the triangleArea() function, and the values of the variables supplied as arguments in the function call are passed as the values of the triangleArea() function's corresponding parameter variables. We can visualize this as

```
area = triangleArea(leg1, leg2);
println("Area: " + area);

}

float hyp(int legA, int legB)
{
    float result;              263        187
    result = sqrt((legA * legA) + (legB * legB));
    return result;
}

float triangleArea(int base, int altitude)
{
    float result;
    result = base * altitude * 0.5;
    return result;
}
```

When the performance of the `triangleArea()` function is complete, the flow of execution returns to the point in the `setup()` program from which the `triangleArea()` function was called. The `float` value resulting from this function call is included in this return. We can visualize this return as

```
    area = triangleArea(leg1, leg2);

    println("Area: " + area);


}

float hyp(int legA, int legB)
{
                                                            24590.5
    float result;
    result = sqrt((legA * legA) + (legB * legB));
    return result;
}

float triangleArea(int base, int altitude)
{
    float result;
    result = base * altitude * 0.5;
    return result;
}
```

Thus, as a result of running the current program, we still see a right triangle of a random size drawn on the canvas:

However, in addition to the leg lengths and the length of the hypotenuse, the *area* of the triangle is now also displayed in the console:

```
leg1: 263
leg2: 187
Hypotenuse: 322.7042
Area: 24590.5
```

As we have seen, we can define and use *multiple* functions in the same program. In this example, we defined and used two functions, but there is no limit to the number of functions that can be defined and used in a program. Both of the two functions defined in this program are functions that return a value. However, we can create multiple void functions in the same program as well. We are also allowed to define both void functions and functions that return a value in the same program.

Save and close the TwoFunctions program.

Another Example: Rounding to Decimal Places

As we know, float values can sometimes be inaccurate, especially in the smaller decimal places. For example, in the case of

```
println(3.0/9.0);
```

Processing produces the following console output:

```
0.33333334
```

This is incorrect. The output should be

```
0.33333333
```

Also, working with so many decimal places is sometimes undesirable. For example, in the case of monetary figures, we might like to work only with two decimal places.* To address both of these kinds of problems, it would be useful to have a function that rounds a float value to a certain number of decimal places.

Select **File > New** to start a new program. Save this program as **Rounding**.

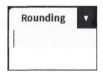

Any given number with decimal places (e.g., 3.168) can be rounded to *tenths* by doing the following:

1) multiply the number by 10	31.68
2) round the result to an integer	32
3) divide the integer by 10	3.2

---

* Actually, because of the inaccuracy of float values, it sometimes advisable to use int values to store monetary amounts. For example, when doing arithmetic with U.S. dollars, it may prove to be more accurate to work with quantities of cents so that int values can be used, rather than using float values to represent fractional dollar amounts.

This formula can be written in Processing code as

```
round(numberToRound * 10) / 10.0
```

Notice that we divide by the `float` value `10.0` (rather than by `10`) in the above formula in order to avoid integer division.

Similarly, given number with decimal places (e.g., 3.168) can be rounded to *hundredths* by doing the following:

1) multiply the number by 100	316.8
2) round the result to an integer	317
3) divide the integer by 100	3.17

This formula could be written in Processing code as

```
round(numberToRound * 100) / 100.0
```

Notice again that we divide by a `float` value, `100.0` (rather than by `100`), in order to avoid integer division.

Now, let's make our formula more general, so that we can round whatever number of decimal places we specify. If *n* is the number of decimal places to which we wish to round, then the more general formula can be described as

1) multiply the number by 10^n

2) round the result to an integer

3) divide the integer by 10^n

For example, to round 3.168 to *two* decimal places, we would do the following:

1) multiply the number by 10^2	316.8
2) round the result to an integer	317
3) divide the integer by 10^2	3.17

We can write this more general formula in Processing code as

```
round(numberToRound * pow(10, places)) / pow(10, places)
```

Because the `pow()` function always returns a `float` value, we do not need to worry about integer division in the above formula.

Let's write a function to round a given `float` value to a specified number of decimal places. As always, when defining a function, we need to supply three items of information in the function *header*:

1) The *type of value to return*.

Because the rounded value that we will be returning will also contain one or more decimal places, we'll define the return type of this function to be `float`.

```
float
```

2) The function's *name*.

If we give this function the name `roundPlaces()`, our header becomes

```
float roundPlaces
```

3) The list of *parameters*.

When we call the `roundPlaces()` function, we need to supply the `float` value that is to be rounded to tenths. Thus, we'll declare a parameter variable to store this supplied value:

```
float roundPlaces(float numberToRound)
```

When we call the `roundPlaces()` function, we need to specify the number of decimal places to which we wish to round the `float` argument supplied. Thus, we'll also need to declare a parameter variable of type `int` to store the number of decimal places specified:

```
float roundPlaces(float numberToRound, int places)
```

With the header of the `roundPlaces()` function complete, we can now begin writing the *body* of the function:

```
float roundPlaces(float numberToRound, int places)
{

}
```

As usual, we will declare a variable named `result` to store the resulting value that this function will return. Because the `roundPlaces()` function is defined to return a value of type `float`, we will declare this variable to be of type `float` as well:

```
float roundPlaces(float numberToRound, int places)
{
  float result;
}
```

Recall, our formula written in Processing code is

```
round(numberToRound * pow(10, places)) / pow(10, places)
```

We will assign the result of this formula to our `result` variable,

```
result = round(numberToRound * pow(10, places)) / pow(10, places);
```

We also need to return this result:

```
return result;
```

Our function definition is now complete and consists of the following:

```
float roundPlaces(float numberToRound, int places)
{
  float result;
  result = round(numberToRound * pow(10, places)) / pow(10, places);
  return result;
}
```

We can now call our `roundPlaces()` function from the `setup()` function. Let's first declare and initialize a variable to hold an unrounded `float` value (e.g., 3.168):

```
void setup()
{
  float unrounded = 3.168;
}
```

Let's also declare and initialize an `int` variable to hold the desired number of decimal places (e.g., two):

```
void setup()
{
  float unrounded = 3.168;
  int decimals = 2;
}
```

We can now use these two variables to create a call to our `floatPlaces()` function. We will use the value of our `unrounded` variable as the `float` argument for the value to be rounded, and we'll use the value of the `decimals` variable as the `int` argument to specify the number of decimal places:

```
floatPlaces(unrounded, decimals)
```

Because our `roundPlaces()` function returns a `float` value when it is called, we need to make immediate use of this returned value. As usual, we'll define a variable for this purpose and assign to this variable the value that is returned by the call to the round-Places() function:

```
void setup()
{
  float unrounded = 3.168;
  int decimals = 2;
  float rounded;
  rounded = roundPlaces(unrounded, decimals);
}
```

Notice that the argument list in the call to the `roundPlaces()` function matches the parameter list in the function definition in terms of the *number* of items, the *types* of items, and the *order* of the items in each list:

```
void setup()
{
  float unrounded = 3.168;
  int decimals = 2;
  float rounded;
  rounded = roundPlaces (unrounded, decimals);
}


float roundPlaces (float numberToRound, int places)
{
  float result;
  result = round(numberToRound * pow(10, places)) / pow(10, places);
  return result;
}
```

Lastly, let's add a statement to display both the unrounded and rounded values to the console:

```
void setup()
{
  float unrounded = 3.168;
  int decimals = 2;
  float rounded;
  rounded = roundPlaces(unrounded, decimals);

  println(unrounded + " --> " + rounded);
}
```

Our complete program is now

```
void setup()
{
  float unrounded = 3.168;
  int decimals = 2;
  float rounded;

  rounded = roundPlaces(unrounded, decimals);

  println(unrounded + " --> " + rounded);
}

float roundPlaces(float numberToRound, int places)
{
  float result;
  result = round(numberToRound * pow(10, places)) / pow(10, places);
  return result;
}
```

When we run this program, both the unrounded `float` value and the same value rounded to the specified number of decimal places are displayed in the console:

```
3.168 --> 3.17
```

If we wish to make this program interactive, so that the user can enter the number to be rounded and the number of decimal places, we can add

```
import static javax.swing.JOptionPane.*;

void setup()
{
  String input;
  String output;
  float unrounded;
  int decimals;
  float rounded;

  char rightArrow = char(8594);

  input = showInputDialog("Enter a number to be rounded:");
  unrounded = float(input);

  input = showInputDialog("Enter a number of decimal places");
  decimals = int(input);

  rounded = roundPlaces(unrounded, decimals);

  output = unrounded + " " + rightArrow + " " + rounded;
  showMessageDialog(frame, output);
}
```

```
float roundPlaces(float numberToRound, int places)
{
  float result;
  result = round(numberToRound * pow(10, places)) / pow(10, places);
  return result;
}
```

When we run this program, the user enters the number to be rounded,

and the number of decimal places:

As a result, both the unrounded and rounded values are displayed to the user in a dialog:

Summary

In this chapter, we learned about creating our own non-void functions, that is, functions that return a value.

- Functions that return a value are known simply as "functions."

- The definition of a function that returns a value requires a return type other than void.

- The definition of a function that returns a value requires a return statement.

- Like a void function, any function that we define to return a value can be called multiple times in the same program.

- Like a void function, any function that returns a value can be defined to have one or more parameters.

- We can define and use multiple void or non-void functions in the same program.

Exercises

1) Write a function named randomMonth() that returns a random integer from 1 to 12. Call the randomMonth() function from the setup() function and assign the returned result to a variable named month. Display the value of month in the console.

   Sample run

   ```
   month number: 7
   ```

2) Write a function named circum () with a float parameter named diameter. Use this parameter along with Processing's built-in PI constant to calculate and return the circumference of a circle that has a diameter equal to diameter. The formula for the circumference of a circle with diameter d is

$$C = \pi d$$

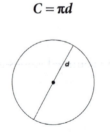

   In the setup() function:

   a) Declare a variable named diam and assign it an int value.

   b) Call the circum () function using diam as an argument and assign the returned result to a variable named circumference.

   c) Display both the value of diam and the value of circum in the console.

   Sample run

   ```
   diameter: 10
   circumference: 31.415928
   ```

3) Modify the program in the preceding exercise to pick a diameter from 1 to 100 at random.

4) Write a function named `circleArea()` with a `float` parameter named `circleRadius`. Use this parameter along with Processing's `PI` constant to calculate and return the area of a circle that has a radius equal to `circleRadius`. The formula for the area of a circle with radius *r* is

$$A = \pi r^2$$

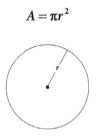

From the `setup()` function,

a) Declare a variable named `radius` and assign it a `float` value.

b) Call the `circleArea()` function using `radius` as an argument and assign the returned result to a variable named `area`.

c) Display both the value of `radius` and the value of `area` in the console.

Sample run

```
radius: 10.0
circle area: 314.15927
```

5) Modify the program in the preceding exercise to pick a diameter from 1 to 100 at random.

6) Write a function named `sphereVolume()` with a `float` parameter named `sphereRadius`. Use this parameter along with Processing's `PI` constant to calculate and return the volume of a sphere with a radius equal to `sphereRadius`. The formula for the volume of a sphere with radius *r* is

$$V = \frac{4}{3}\pi r^3$$

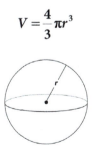

(Hint: Watch out for integer division.)

In the setup() function,

a) Declare a variable named radius and assign it a float value.

b) Call the sphereVolume() function using radius as an argument and assign the returned result to a variable named volume.

c) Display both the value of radius and the value of volume in the console.

Sample run

```
radius: 10.0
sphere volume: 4188.79
```

7) Modify the program in the preceding exercise to pick a diameter from 1 to 100 at random.

8) Write a function named coneVolume () with float parameters named coneRadius and coneAltitude. Use these parameters along with Processing's PI constant to calculate and return the volume of a cone with a radius equal to coneRadius and an altitude equal to coneAltitude. The formula for the volume of a cone with a radius r and altitude a is

$$V = \frac{1}{3}\pi r^2 a$$

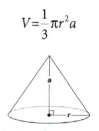

(Hint: Watch out for integer division.)
In the setup() function,

a) Declare a variable named radius and assign it a float value.

b) Declare a variable named altitude and assign it a float value.

c) Call the coneVolume() function using radius and altitude as arguments and assign the returned result to a variable named volume.

d) Display the value of radius, the value of altitude, and the value of volume in the console.

Sample run

```
radius: 10.0
altitude: 20.0
cone volume: 2094.395
```

9) Write a function named yardsToMeters() that receives a number of yards as a float argument and returns the equivalent number of meters (1 yard = 0.9144 meters).

In the setup() function,

a) Pick a random integer between 1 and 100 as a number of yards and assign this to an int variable named yards.

b) Call the yardsToMeters() function using yards as an argument and assign the returned result to a variable named meters.

c) Display the selected number of yards and the equivalent number of meters on a single line in the console.

Sample run

```
32 yards is equivalent to 29.2608 meters.
```

10) Modify the program in the preceding exercise to allow the user to enter in a dialog box the number of yards to be converted to meters.

11) Write a function named monthDays () that receives a month number and returns a String value indicating the number of days in that month. Use the if-else if-else or switch statement example from Chapter 4 to determine the number of days for a given month.

In the setup() function,

a) Pick a random integer from 1 to 12.

b) Call the monthDays() function with the selected random integer as an argument.

c) Display both the selected month number and the number of days in the console.

Sample runs

```
Month 5 has 31 days.
Month 9 has 30 days.
Month 2 has 28 or 29 days.
```

12) Modify the program in the preceding exercise to allow the user to enter the month number in a dialog box.

CHAPTER 9

Arrays

Consider the following list of six temperatures:

57, 10, 32, 78, 212, 101

Suppose we need to work with all six of these temperatures in a single computer program. One way to do so would be to declare a variable for each of them:

```
int temperature1 = 57;
int temperature2 = 10;
int temperature3 = 32;
int temperature4 = 78;
int temperature5 = 212;
int temperature6 = 101;
```

However, this seems to be a somewhat cumbersome approach. Think of how unwieldy this approach would be if we needed to process 100 temperatures!

Fortunately, there is a better way to work with lists of related elements that are all the same type: by using an *array* of values.

About Arrays

Up to now, each of the variables we have used has stored a *single* value. However, like most programming languages, Processing also provides us with the ability to use an *array* of values.

An **array** consists of a fixed number of items of the same type arranged in a specific order.

Let's look more closely at this definition:

1) "fixed number"

An array is always defined to store a specific number of items, and this storage capacity does not change.*

2) "same type"

All the values stored in an array must be of the same type.

3) "specific order"

The items in an array are kept in a specific order.

Each item stored in an array is called an **element**.

To illustrate an array, suppose we were to insert our list of temperatures into the upper-left corner of a spreadsheet and look only at the *row numbers* of these temperatures:

1	57
2	10
3	32
4	78
5	212
6	101

This is a good visual image for thinking about arrays. There is one key difference, however. In computer programming, the numbering of items often starts at *zero* rather than at 1. For example, we have seen this in the numbering of pixel rows and pixel columns of the canvas. We see such zero-based numbering in arrays as well. Here, we need to think of the six temperatures in our array as numbered from 0 to 5 (not 1 to 6):

0	57
1	10
2	32
3	78
4	212
5	101

An array always has a *name*. Let's give the name "temperatures" to this array.

temperatures	
0	57
1	10
2	32
3	78
4	212
5	101

---

* Technically, Processing does include functions for expanding and shortening an array. However, these functions are not typical of the way that arrays are used in most other programming languages.

Now, suppose that we want to refer to a particular element in our array. In Processing, the way to do this is simply by giving the name of the array, followed by the desired row number, enclosed in square brackets. The row number that refers to a particular element in an array is called an **index** or **subscript**. For example, we can refer to the element in row 3 of the temperatures array as

    temperatures[3]

temperatures	
0	57
1	10
2	32
3	78
4	212
5	101

Using our array terminology, we can say that this element in the temperatures array has an index (or subscript) of 3. Likewise, we can say that the element currently stored in temperatures[3] is 78. Notice that this element has an index of 3, but it is the *fourth* element in the array because the first index of any array is *zero*.

Declaring an Array

Let's actually create the above temperatures array as a variable in Processing. Select **File > New** to start a new program. Save this program as **TempsArray**.

The first step in creating an array variable is *declaring* it. As usual when declaring a variable, we begin the declaration for an array variable by specifying a *type*:

    int

However, when we declare an *array* variable, we are specifying the type of *all* the elements that will be stored in the array. For this reason, when specifying the type of an array variable, we add left and right *square brackets* after the type specifier:

    int[]

Next, we specify the *name* of our array.* This declaration of our array variable is now complete:

```
int[] temperatures;
```

However, the above statement merely instructs Processing to create a variable named temperatures that is able to store the *address* in the computer's memory where an array of int values will be located. This array has not actually been created yet, so the memory address is not yet known. We can visualize this as

int[] | **????**
temperatures

In order to create an array in the computer's memory, Processing needs to know how many elements our array will have, so that it can set aside sufficient space for this array in memory.

Save the TempsArray program. We will continue with it in the next section.

Initializing an Array with Specific Values

Return to the TempsArray program. Sometimes, we already have in mind the initial values that we would like our new array to contain when we create it. In such cases, we can insert these values to the array variable using the assignment operator (=) within our array declaration statement. We simply list the values, separated by commas, and enclose this list between left and right curly braces. For example, we can both declare the temperatures array and initialize it to contain the six specific int values in our above example using the following statement:

```
int[] temperatures = {57, 10, 32, 78, 212, 101};
```

However, note that we are *not* able to declare the array variable and specify initial values in *separate* statements.† For example, the following two statements generate an error:

```
int[] temperatures;
temperatures = {57, 10, 32, 78, 212, 101};←error!
```

When we declare and initialize our array in a single statement like this, Processing does *three* things:

---

* Although it is a less common practice, we are also allowed to put the square brackets after the array name instead of after the type, for example,

```
int temperatures[] = {57, 10, 32, 78, 212, 101};.
```

† Technically, declaring and initializing an array in two separate statements is possible, but this technique is quite unconventional:

```
int[] temperatures;
temperatures = new int[]{57, 10, 32, 78, 212, 101};
```

1) Processing sets aside space for this new array in the computer's memory.

   In this particular case, Processing determines that we have six elements of type `int`, so it sets aside sufficient space in memory for six elements of this type.

2) Processing inserts the values we have listed into the corresponding locations in the array.

3) Processing stores in the array variable the numeric *address* in the computer's memory where this new array is located.

Using our `temperatures` array example, we can visualize the resulting relationship between our array variable and the array's location in memory as

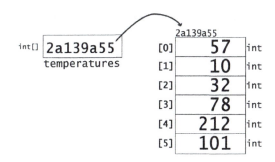

Thus, an array variable does *not* contain the actual array. Rather, an array variable contains the numeric *address* in the computer's memory where the array is stored. As evidence of this, we can call the `print()` function using our array variable:

```
int[] temperatures = {57, 10, 32, 78, 212, 101};
print(temperatures);
```

When we run this program, we see console output something like the following:

```
[I@2a139a55
```

The somewhat odd-looking sequence of characters you see after the @ symbol describes the numeric address in memory where the array is located.* These are "hexadecimal" (base-16) digits, which allow for a more compact way of writing the large numbers needed for numbering memory addresses.†

---

* The address you see will be different from the one shown here.
† Today's computers have gigabytes of memory, so numbers in the billions are required to assign a unique numeric address to each location of a computer's memory. For example, 705,927,765 is the decimal equivalent of 2a139a55, the hexadecimal (base-16) number shown here. However, the numeric addresses are binary numbers. For example, the number shown requires 30 binary digits: 101010000100111001101001010101. This is why hexadecimal numbers are used as a kind of "shorthand" version of binary numbers: because $2^4 = 16$, every four binary (base-2) digits can be compacted into one hexadecimal (base-16) digit. Thus, the 20-digit binary number can be compacted into the 8-digit hexadecimal number.

For certain uses of arrays, it's important that we understand the behind-the-scenes relationship between an array variable and the actual array that is elsewhere in the computer's memory. However, for most of our uses of arrays, we can visualize an array as simply

temperatures	
0	57
1	10
2	32
3	78
4	212
5	101

Save the `TempsArray` program. We will continue with it in the next section.

Displaying an Array: The `printArray()` Function

Return to the `TempsArray` program. We can easily test whether our array has been successfully storing our desired values using Processing's **`printArray()`** function.* Let's make the following change to our program:

```
int[] temperatures = {57, 10, 32, 78, 212, 101};
printArray(temperatures);
```

When we run this program, we now see the following output to the console:

```
[0] 57
[1] 10
[2] 32
[3] 78
[4] 212
[5] 101
```

Save the `TempsArray` program. We will continue working with it in the next section.

Accessing an Array Element

Return to the `TempsArray` program. We are *not* required to process the elements of an array *in sequence*. Rather, an array provides **random access**, which means that any of the array locations is directly accessible. To access any array element, we simply use the appropriate index. Thus, given the array declaration

```
int[] temperatures = {57, 10, 32, 78, 212, 101};
```

we can refer to the first element in the array as

```
temperatures[0]
```

---

* As of this writing, the `println()` function also still works to display an array in this same format, but the newer `printArray()` function is the one recommended for doing so.

For example, to display the first element of the `temperatures` array, we can make the following change to our program:

```
int[] temperatures = {57, 10, 32, 78, 212, 101};
println(temperatures[0]);
```

This causes a copy of the first element in this array to be retrieved from memory and used as an argument in the call to the `println()` function.

temperatures	
0	57
1	10
2	32
3	78
4	212
5	101

```
57
println(temperatures[0]);
```

Thus, when we run this program, we see the following output to the console:

```
57
```

Save the `TempsArray` program. We will continue working with it in the next section.

Storing a Value in an Array

Return to the `TempsArray` program. To *store* a value in an array element, we simply use an assignment statement.

For example, if we wish to change the first element in the array to 62, we can insert the following assignment statement into our program:

```
int[] temperatures = {57, 10, 32, 78, 212, 101};

temperatures[0] = 62;

printArray(temperatures[0]);
```

This causes the `int` value 62 to be stored as the first element in the array.

```
temperatures[0] = 62;
```

temperatures	
0	62
1	10
2	32
3	78
4	212
5	101

Let's now modify the third statement in our program so that it displays all the array elements in the console:

```
int[] temperatures = {57, 10, 32, 78, 212, 101};

temperatures[0] = 62;

printArray(temperatures);
```

When we run this program, we see the following output to the console, which shows that the value 62 is indeed now stored as the first element of the array:

```
[0] 62
[1] 10
[2] 32
[3] 78
[4] 212
[5] 101
```

Save and close the TempsArray program.

Processing an Array

Counting loops are excellent tools for working with arrays. The for statement in particular is commonly used when it is necessary to step through the elements of an array.

Select **File > New** to start a new program. Save this program as **ProcessArray**.

ProcessArray

Let's start again with the following temperatures array:

```
int[] temperatures = {57, 10, 32, 78, 212, 101};
```

We can use a for statement to display all the values in this array. For this, we need a for loop that will count through all the indexes in the temperatures array, from 0 up to 5. Thus, the heading of our for statement can be

```
int[] temperatures = {57, 10, 32, 78, 212, 101} ;

for (int count = 0; count <= 5; count = count + 1)
{

}
```

To clarify the looping action of this for statement, let's insert the following call to the println() function into the body of this statement:

```
int[] temperatures = {57, 10, 32, 78, 212, 101};

for (int count = 0; count <= 5; count = count + 1)
{
  println(count);
}
```

When we run this program, we see from the console output that the counting variable in this for statement, count, is counting from 0 to 5:

```
0
1
2
3
4
5
```

Thus, we can modify the call to the println() function so that count is instead used as the *index* of an element in the temperatures array:

```
int[] temperatures = {57, 10, 32, 78, 212, 101};

for (int count = 0; count <= 5; count = count + 1)
{
  println(temperatures[count]);
}
```

Now, when we run our program, we see that all the elements of the temperatures array are displayed in the console, in order:

```
57
10
32
78
212
101
```

Let's take a closer look at how this for statement achieves this result. The *first* time the body of this for statement is performed, the value of count is 0, so temperatures [0] is displayed in the console. We can visualize this as

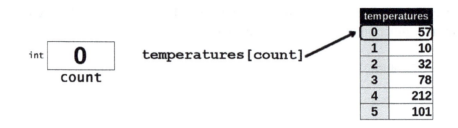

Thus, the console output produced is

57

The *second* time the body of this `for` statement is performed, the value of `count` is 1, so `temperatures [1]` is displayed in the console. We can visualize this as

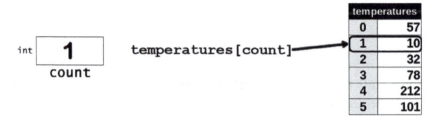

Thus, the console output produced is

10

The *third* time the body of this `for` statement is performed, the value of `count` is 2, so `temperatures [2]` is displayed in the console. We can visualize this as

Thus, the console output produced is

32

The *fourth* time the body of this `for` statement is performed, the value of `count` is 3, so `temperatures [3]` is displayed in the console. We can visualize this as

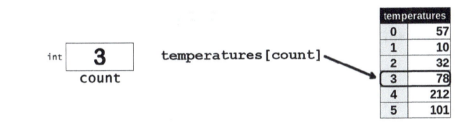

Thus, the console output produced is

78

The *fifth* time the body of this for statement is performed, the value of count is 4, so temperatures [4] is displayed in the console. We can visualize this as

Thus, the console output produced is

212

The *fifth* and final time the body of this for statement is performed, the value of count is 5, so temperatures [5] is displayed in the console. We can visualize this as

Thus, the console output produced is

101

As we've seen, the overall action of this for statement is stepping through the entire array, one element at a time, in order. We can visualize this as

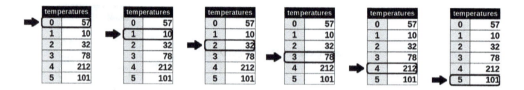

And the total console output produced by this `for` statement is

```
57
10
32
78
212
101
```

By using a `for` statement to display the elements of the array, we gain more *flexibility* than if we use the `printArray()` function. For example, let's change the call to the `println()` function in the body of our `for` statement to the following:

```
int[] temperatures = {57, 10, 32, 78, 212, 101};

for (int count = 0; count <= 5; count = count + 1)
{
    println("temperatures[" + count + "] = " + Temperatures[count]);
}
```

Now, when we run this program, we see the following output to the console:

```
temperatures[0] = 57
temperatures[1] = 10
temperatures[2] = 32
temperatures[3] = 78
temperatures[4] = 212
temperatures[5] = 101
```

Save the `ProcessArray` program. We will continue working with it in the next section.

Choosing Loop Conditions Carefully When Processing an Array

We need to set the loop condition *carefully* when using a counting loop with an array. For example, because we think of the `temperatures` array as holding six elements, we might be inclined to write the following:

```
int[] temperatures = {57, 10, 32, 78, 212, 101};
```

```
for (int count = 0; count <= 6; count = count + 1)
{
  println("temperatures[" + count + "] = " + temperatures[count]);
}
```

We have now created a for loop that will count from 0 up to 6, whereas the indexes of our array run only from 0 to 5. Thus, when we run our program, Processing does display the six elements of the temperatures array in the console:

```
temperatures[0] = 57
temperatures[1] = 10
temperatures[2] = 32
temperatures[3] = 78
temperatures[4] = 212
temperatures[5] = 101
```

However, Processing then displays an error message:

ArrayIndexOutOfBoundsException: 6

This error message alerts us that an array index has gone out of bounds when Processing tried to use an array index of 6. We can visualize this as

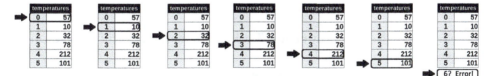

In other words, when the loop counter variable reached 6, Processing tried to output to the console the value of temperatures[6]. However, there is no seventh element in the array with an index of 6, so Processing stopped the program and displayed the above error message.* Because the temperatures array can hold six elements, we say that the **length** of this array is 6. However, it is important to remember that although the length of the array is 6, the *last* item in this array is not temperatures[6], but rather temperatures[5]. This is because the index of the *first* element in an array is always *zero*. **The index of the *last* item in an array is always *one less than* the *length* of that array.** Save the ProcessArray program. We will continue working with it in the next section.

The length Property of an Array

Return to the ProcessArray program. A good way to avoid the problem of an array index going out of bounds is by remembering that the index of an array must always

---

* An **exception** is a computer programming term that means something goes wrong when we try to run our program.

remain strictly *less than* the *length* of the array. For this reason, given that the length of our `temperatures` array is 6, a better way to write the header of our `for` statement might be

```
int[] temperatures = {57, 10, 32, 78, 212, 101};

for (int count = 0; count < 6; count = count + 1)
{
  println("temperatures[" + count + "] = " + temperatures[count]);
}
```

This version of the loop condition still ensures that the loop counter variable will only count from 0 to 5.

However, what if we later decide to use *more* or *fewer* elements in our array? This means that our current `for` statement will no longer work correctly to process this array. For example, suppose we would like to increase the number of our `temperature` array's initial values by 1:

```
int[] temperatures = {57, 10, 32, 78, 212, 101, 15};

for (int count = 0; count < 6; count = count + 1)
{
  println("temperatures[" + count + "] = " + temperatures[count]);
}
```

This brings the length of the array up to 7. The index of the last element is 6. We can visualize this as

temperatures	
0	57
1	10
2	32
3	78
4	212
5	101
6	15

However, when we run this program, our `for` loop counts only from 0 to 5. Thus, only the first six elements of the array are displayed to the console, instead of all seven:

```
temperatures[0] = 57
temperatures[1] = 10
temperatures[2] = 32
temperatures[3] = 78
temperatures[4] = 212
temperatures[5] = 101
```

Or, suppose that we *shorten* the list of initial elements:

```
int [] temperatures = {57, 10, 32, 78, 212};

for (int count = 0; count < 6; count = count + 1)
{
  println("temperatures[" + count + "] = " + temperatures[count]);
}
```

This brings the length of the array down to 5. The index of the last element is 4. We can visualize this as

temperatures	
0	57
1	10
2	32
3	78
4	212

When we run our program, the `for` loop now counts from 0 to 4. Thus, the first five elements of the array are displayed in the console:

```
temperatures[0] = 15
temperatures[1] = 57
temperatures[2] = 10
temperatures[3] = 32
temperatures[4] = 78
```

However, afterward, the `for` loop goes on to count to 5, and we once again receive an error message notifying us that an array index is out of bounds:

ArrayIndexOutOfBoundsException: 5

This exception occurs because there currently is no element in the `temperatures` array that has an index of 5. The length of our array is 5, so the index of the last element is one less than this length: 4.

Let's reinsert the sixth element we have been using in our array:

```
int [] temperatures = {57, 10, 32, 78, 212, 101};

for (int count = 0; count < 6; count = count + 1)
{
  println("temperatures[" + count + "] = " + temperatures[count]);
}
```

The length of our array is once again 6. The index of the last element is 5. We can visualize this as

Thus, when we run our program, we once again see that our `for` loop correctly displays all the elements of our array:

```
temperatures[0] = 57
temperatures[1] = 10
temperatures[2] = 32
temperatures[3] = 78
temperatures[4] = 212
temperatures[5] = 101
```

Fortunately, there is a *better* way for us to construct the heading of a `for` statement using specific values for the maximum index and running the risk of missing the last array element or having an array index go out of bounds. One of the identifiable features of an array as an *object* is the built-in **length** property that every array has. This `length` property of an array keeps track of an array's length—that is, its capacity, how many elements that array can hold. To access an array's `length` property, we simply write the name of the array, followed by a period (dot) and then the keyword `length`:

> *array-name*.**length**

(This is another example of what is known as *dot notation*.) For example, given the following `temperatures` array, we can display the length of this array by adding the following statement to our program:

```
int[] temperatures = {57, 10, 32, 78, 212, 101};
for (int count = 0; count < 6; count = count + 1)
{
  println("temperatures[" + count + "] = " + temperatures[count]);
}

println("array length: " + temperatures.length);
```

When we run this program, we see the following output:

```
temperatures[0] = 57
temperatures[1] = 10
temperatures[2] = 32
temperatures[3] = 78
temperatures[4] = 212
temperatures[5] = 101
array length: 6
```

On the last line of the console output, we see the value of the temperature array's length property displayed: 6. This indicates that our temperature array currently has six elements.

An array's length property is especially useful when creating the condition of a counting loop to process an array. Remember that the index of the last element of an array is always *one less than* the length of that array. Thus, we can modify the condition of the for statement in our current program to the following:

```
int[] temperatures = {57, 10, 32, 78, 212, 101};

for (int count = 0; count < temperatures.length; count = count + 1)
{
  println("temperatures[" + count + "] = " + temperatures[count]);
}

println("array length: " + temperatures.length);
```

When we run our program, we see the same output as before:

```
temperatures[0] = 57
temperatures[1] = 10
temperatures[2] = 32
temperatures[3] = 78
temperatures[4] = 212
temperatures[5] = 101
array length: 6
```

Using an array's length property makes our code more *flexible*. As we have seen, if we use a *specific* value in our loop condition and then later *increase* or *decrease* the number of values in the list we use to initialize our array, then our for loop will no longer work correctly to process the array. In contrast, if we use the length property in our for statement, we ensure that our for loop will count through all the array elements, regardless of how many elements there are in the array. For example, suppose we *add* an element to our original list of the array's initial values:

```
int[] temperatures = {57, 10, 32, 78, 212, 101, 15};
for (int count = 0; count < temperatures.length; count = count + 1)
{
```

```
  println("temperatures[" + count + "] = " + temperatures[count]);
}

println("array length: " + temperatures.length);
```

This brings the number of array elements up to seven. The index of the last element is 6. We can visualize this as

temperatures	
0	57
1	10
2	32
3	78
4	212
5	101
6	15

By using the array's length property, when we run this program, all seven elements are displayed to the console, along with the array's new length:

```
temperatures[0] = 57
temperatures[1] = 10
temperatures[2] = 32
temperatures[3] = 78
temperatures[4] = 212
temperatures[5] = 101
temperatures[6] = 15
array length: 7
```

Likewise, suppose that we *shorten* the list to only five initial elements:

```
int[] temperatures = {57, 10, 32, 78, 212};

for (int count = 0; count < temperatures.length; count = count + 1)
{
  println("temperatures[" + count + "] = " + temperatures[count]);
}
println("array length: " + temperatures.length);
```

The length of the array is now 5. The index of the last element in the array is 4. We can visualize this as

temperatures	
0	57
1	10
2	32
3	78
4	212

By using the array's `length` property, when we run the preceding program, all five elements of the array are displayed in the console, along with the array's new list, without an error message resulting:

```
temperatures[0] = 57
temperatures[1] = 10
temperatures[2] = 32
temperatures[3] = 78
temperatures[4] = 212
array length: 5
```

Finally, suppose we return to our original list of six elements:

```
int[] temperatures = {57, 10, 32, 78, 212, 101};

for (int count = 0; count < 6; count = count + 1)
{
  println("temperatures[" + count + "] = " + temperatures[count]);
}

println("array length: " + temperatures.length);
```

The length of the array is once again 6. The index of the last element in the array is once again 5. We can visualize this as

temperatures	
0	57
1	10
2	32
3	78
4	212
5	101

Thus, when we run our program, we once again see the proper output of the array elements and array length:

```
temperatures[0]  =  57
temperatures[1]  =  10
temperatures[2]  =  32
temperatures[3]  =  78
temperatures[4]  =  212
temperatures[5]  =  101
array length:  6
```

As we can see, by using the `length` property of the array in the condition of our `for` statement, we do not need to make any modifications to the `for` statement if we change the number of initial elements of our array. Using the `length` property in this way also makes our code more *readable*. When we write our `for` statement's test condition as

```
count < temperatures.length
```

we make it clearer to a reader of our code that we wish to stop the value of the loop counter variable from going beyond the end of the array. If we had used a specific value,

```
count < 6
```

then our intention would be less clear.

Save the `ProcessArray` program. We will continue working with it in the next section.

Processing an Array: Calculating the Average Temperature

Let's look at another example of processing an array: calculating an *average* of the elements of the array. Return to the `ProcessArray` program. Select **File > Save As** and resave the `ProcessArray` program as **AverageTemp**.

We can visualize our current array as

temperatures	
0	57
1	10
2	32
3	78
4	212
5	101

And our current program is

```
int[] temperatures = {57, 10, 32, 78, 212, 101};
for (int count = 0; count < temperatures.length; count = count + 1)
{
  println("temperatures[" + count + "] = " + temperatures[count]);
}

println("array length: " + temperatures.length);
```

When we run this program, we see the following output to the console:

```
temperatures[0] = 57
temperatures[1] = 10
temperatures[2] = 32
temperatures[3] = 78
temperatures[4] = 212
temperatures[5] = 101
array length: 6
```

In order to calculate the average temperature, we will first need to calculate the *sum* of the temperatures in our array. Because the values in the temperatures array are of type int, it seems to make sense to use an int variable to calculate this sum. We also need to initialize this variable to zero. To help keep our code more concise, let's perform this declaration and initialization in a single statement by writing the following:

```
int[] temperatures = {57, 10, 32, 78, 212, 101};
int sum = 0;

for (int count = 0; count < temperatures.length; count = count + 1)
{
  println("temperatures[" + count + "] = " + temperatures[count]);
}

println("array length: " + temperatures.length);
```

We can visualize the effect of this added statement as

int **0**

sum

We can easily keep a running total of the elements as we step through the array by inserting the following statement into our for loop:

```
for (int count = 0; count < temperatures.length; count = count + 1)
{
  println("temperatures[" + count + "] = " + temperatures[count]);
  sum = sum + temperatures[count];
}
```

Let's take a closer look at how this assignment statement will operate. As always, any *expression* on the right-hand side of the assignment operator (=) will be evaluated *first*. As before, you may add parentheses to enclose the right-hand side if it makes this assignment statement easier for you to understand:

```
sum = (sum + temperatures[count]);
```

At this point, the values of the sum and count variables are both zero. We can visualize this as

To evaluate this expression on the right-hand side, the current value of the sum variable must be retrieved. During this first pass through the for loop, the initial value retrieved from the sum variable is 0. We can visualize this as

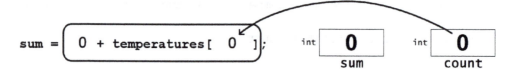

The value of the count variable during the first pass through the for loop is also 0. Thus, our expression essentially becomes

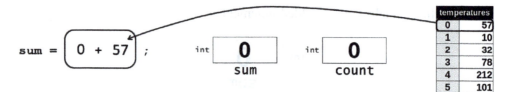

The value retrieved from temperatures[0] is 57. We can visualize this as

sum = | 0 + 57 | ;   int | **0** | int | **0** |

temperatures	
0	57
1	10
2	32
3	78
4	212
5	101

sum   count

Thus, the addition operation essentially becomes

sum = | 0 + 57 | ;

The addition operation is now performed, completing the evaluation of the right-hand side. Thus, the assignment statement becomes

Thus, we can visualize this assignment operation and its result as

During the *second* pass through the `for` loop, this assignment statement is performed again:

```
for (int count = 0; count < temperatures.length; count = count + 1)
{
  println("temperatures[" + count + "] = " + temperatures[count]);
  sum = sum + temperatures[count];
}
```

Once again, the expression on the right-hand side of the assignment operator (=) will be evaluated *first*. The value of our `sum` variable is currently 57, and the value of the `count` variable is currently 1. We can visualize this as

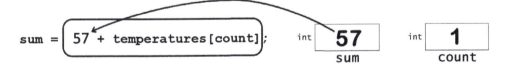

To evaluate this expression on the right-hand side, the current value of the `sum` variable must be retrieved. We can visualize this as

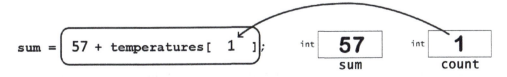

A copy of the current value of the `count` variable is also retrieved. We can visualize this as

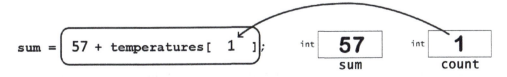

The value retrieved from `temperatures[1]` is 10. We can visualize this as

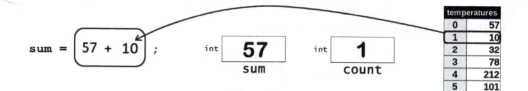

Thus, the addition operation in the expression essentially becomes

The addition operation can now be performed, completing the evaluation of the right-hand side. Thus, the assignment statement becomes

sum = (67) ;

Thus, 67 is stored as the new value of the sum variable. We can visualize this assignment operation and its result as

sum = (67) ;   int 67
                        sum

This assignment statement is performed four more times, once during each of the remaining four passes through the for loop:

```
for (int count = 0; count < temperatures.length; count = count + 1)
{
  println("temperatures[" + count + "] = " + temperatures[count]);
  sum = sum + temperatures[count];
}
```

The final result is that the total of all six elements of the array is stored in the sum variable. We can visualize this as

Let's add a statement after the for loop to display the resulting value of the sum variable, so we can check the sum that has been calculated:

```
int[] temperatures = {57, 10, 32, 78, 212, 101};
int sum = 0;
for (int count = 0; count < temperatures.length; count = count + 1)
{
  println("temperatures[" + count + "] = " + temperatures[count]);
  sum = sum + temperatures[count];
}

println("array length: " + temperatures.length);

println("Sum: " + sum);
```

When we run this program, we now see the following output to the console:

```
temperatures[0] = 57
temperatures[1] = 10
temperatures[2] = 32
temperatures[3] = 78
temperatures[4] = 212
temperatures[5] = 101
array length: 6
Sum: 490
```

Now that we have the sum of our temperatures, we are ready to calculate the average temperature. The resulting average might not be an integer, so let's declare a float variable to store this average:

```
float average;
```

To calculate the average, we divide the sum by the number of elements. Thus, we might be tempted to add the following assignment statement to our program:

```
average = sum / 6;
```

However, what if we later decide to *change* the number of elements in the array—for example, increase the length of the array to 7 or decrease it to 5? Our calculated average will then be incorrect. A better approach is to use our array's length property here:

```
average = sum / temperatures.length;
```

Thus, our program is now

```
int[] temperatures = {57, 10, 32, 78, 212, 101};
```

```
int sum = 0;
float average;

for (int count = 0; count < temperatures.length; count = count + 1)
{
  println("temperatures[" + count + "] = " + temperatures[count]);
  sum = sum + temperatures[count];
}

println("array length: " + temperatures.length);
println("Sum: " + sum);

average = sum / temperatures.length;
println("Average: " + average);
```

However, when we run our program, we see the following output:

```
temperatures[0] = 57
temperatures[1] = 10
temperatures[2] = 32
temperatures[3] = 78
temperatures[4] = 212
temperatures[5] = 101
array length: 6
Sum: 490
Average: 81.0
```

The resulting average is not correct. The correct average should be approximately 81.666667. The problem here is that, as we can see from our console output, the length property of an array is stored as an int value:

```
array length: 6
```

Thus, the expression

```
sum / temperature.length
```

is *integer* division, because both the variable sum and the temperature array's length property are of type int. In this particular case, the current value of sum is 490, and the current value of temperatures.length is 6. Thus, the integer division in the expression

```
sum / temperature.length
```

is essentially

```
490 / 6
```

This integer division operation evaluates to 81, an int value. As we know, an int value is automatically converted by Processing to the float value when we assign it to a float variable. Thus, when the value 81 is assigned to average, it is converted to 81.0, so the resulting value of the average variable is 81.0.

How can we fix this? Remember that if even *one* of the two operands in an arithmetic operation is a float value, then the result of that operation will be a float value. Remember also that the float() function will convert an int value to the equivalent float value. (We can think of this as if Processing simply adds a .0 to the int value.) Thus, we can convert sum to a float value using the float() function by inserting the following:

```
average = float(sum) / temperatures.length;
```

Now, we avoid integer division and will get the correct result. Thus, our program is now

```
int[] temperatures = {57, 10, 32, 78, 212, 101};

int sum = 0;
float average;

for (int count = 0; count < temperatures.length; count = count + 1)
{
  println("temperatures[" + count + "] = " + temperatures[count]);
  sum = sum + temperatures[count];
}

println("array length: " + temperatures.length);
println("Sum: " + sum);

average = float(sum) / temperatures.length;
println("Average: " + average);
```

Now, when we run this program, we see the following output:

```
temperatures[0] = 57
temperatures[1] = 10
temperatures[2] = 32
temperatures[3] = 78
temperatures[4] = 212
temperatures[5] = 101
array length: 6
Sum: 490
Average: 81.666664
```

Notice that the last digit of the average illustrates once again the inherent inaccuracy of the last decimal places on float values when small fractions are involved:

```
Average: 81.666664
```

Remember to be careful when using `float` values.*

Save and close the `AverageTemp` program.

Graphical Example: Graphing an Array

Let's create a program to produce a graph of the values stored in our `temperatures` array:

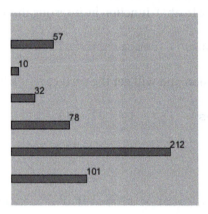

Select **File > New** to start a new program. Save this program as **GraphTemps**.

```
GraphTemps ▼

|
```

We'll need the array of temperatures that we have been using in this chapter:

```
int[] temperatures = {57, 10, 32, 78, 212, 101};
```

To match the graph shown above, we'll set the canvas size to 250 pixels by 250 pixels:

```
size(250, 250);
```

In the graph shown above, the distance from the top of one bar to the top of the next bar beneath it is 35 pixels:

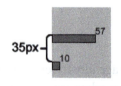

---

* We could use the `roundPlaces()` function from Chapter 8 to round this function to fewer decimal places.

Thus, let's create a variable to store this value:

```
int spaceBetween = 35;
```

Also, the height of each bar is 10 pixels:

Let's also create a variable to store this value:

```
int barHeight = 10;
```

Thus, our program is now

```
int[] temperatures = {57, 10, 32, 78, 212, 101};

size(250, 250);

int spaceBetween = 35;
int barHeight = 10;
```

We will need to calculate the pixel *row* location of the *top* of each bar that we draw. Let's declare a variable to store this value as well:

```
size(250, 250);

int spaceBetween = 35;
int barHeight = 10;
int barTop;
```

To count through the temperatures in our array, we need a `for` statement:

```
size(250, 250);

int spaceBetween = 35;
int barHeight = 10;
int barTop;

for (int count = 0; count < temperatures.length; count = count + 1)
{
}
```

Next, we need to insert into the body of this `for` loop the statements that are needed to draw the rectangle for each bar. We'll start by setting the fill color to red:

```
for (int count = 0; count < temperatures.length; count = count + 1)
{
  fill(255, 0, 0);// red
}
```

Next, we need to draw the rectangle. For this, we will need to know the pixel *row* location of the rectangle's upper-left corner. This depends on which bar we are drawing, so we'll need to calculate this row position. We know that the space between the bars is 35 pixels:

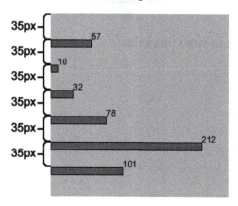

Thus, as our loop counter variable counts from 0 to 5, we can describe the pixel row position of each bar as

```
(count + 1) * 35
```

However, because we stored this value in our spaceBetween variable, we can instead write our expression as

```
(count + 1) * spaceBetween
```

As we count through our for loop, this expression will calculate the correct pixel row number using each value of count, from 0 to 5. We declared the variable barTop so that we can store this calculated pixel row position. Thus, we can assign the result of the above expression to this variable with the following statement:

```
for (int count = 0; count < temperatures.length; count = count + 1)
{
  fill(255, 0, 0);// red
  barTop = (count + 1) * spaceBetween;
}
```

We are now ready to call the rect() function. The first argument needed for this function is the pixel *column* position of the upper-left corner of the rectangle. For each of the bars, this will always be at column 0, the left edge of the canvas. Thus, we start our call to the rect() function with

```
for (int count = 0; count < temperatures.length; count = count + 1)
{
  fill(255, 0, 0);// red
  barTop = (count + 1) * spaceBetween;
  rect(0,
}
```

The next argument of the rect() function needed is the pixel *row* location of the upper-left corner of the rectangle. We have stored this calculated value in our barTop variable. Our call to the rect() function is now

```
for (int count = 0; count < temperatures.length; count = count + 1)
{
  fill(255, 0, 0);// red
  barTop = (count + 1) * spaceBetween;
  rect(0, barTop,
}
```

The next argument for the rect() function is the *width* of the rectangle. Because the length of each bar is 1 pixel for each degree, the width of the rectangle will always be the same number of pixels as the value of the current array element. For example, the value of temperatures[0] is 57. We want the width our rectangle to be the same: 57 pixels. Thus, our call to the rect() function is now

```
for (int count = 0; count < temperatures.length; count = count + 1)
{
  fill(255, 0, 0);// red
  barTop = (count + 1) * spaceBetween;
  rect(0, barTop, temperatures[count],
}
```

The final argument needed for the rect() function is the desired *height* of the rectangle. We have already stored this value in our barHeight variable. Thus, we can complete our call to the rect() function:

```
for (int count = 0; count < temperatures.length; count = count + 1)
{
  fill(255, 0, 0);// red
  barTop = (count + 1) * spaceBetween;
  rect(0, barTop, temperatures[count], barHeight);
}
```

Our program is now

```
int[] temperatures = {57, 10, 32, 78, 212, 101};
```

```
size(250, 250);

int spacer = 35;
int barHeight = 10;
int barTop;

for (int count = 0; count < temperatures.length; count = count + 1)
{
  fill(255, 0, 0);// red
  barTop = (count + 1) * spacer;
  rect(0, barTop, temperatures[count], barHeight );
}
```

When we run this program, we do see all six of the bars drawn correctly:

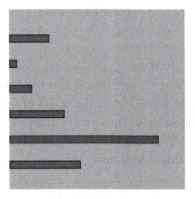

Next, we need to add black text at the end of each rectangle. Recall that the fill() function is used to set the text color generated by the text() function. Thus, we add a call to the fill() function inside of our for statement:

```
for (int count = 0; count < temperatures.length; count = count + 1)
{
  fill(255, 0, 0);// red
  barTop = (count + 1) * spacer;
  rect(0, barTop, temperatures[count], barHeight);
  fill(0, 0, 0);// black
}
```

Recall that the basic form of a call to the text() function is

```
text(text-to-display, column, row);
```

The first argument of the text() function, the text that we wish to display on the canvas, is the value of the current array element. Thus, we begin the call to the text() function with

```
text(temperatures[count],
```

The second argument needed for the `text()` function will specify the horizontal pixel location of our text. We want our text to be positioned at the end of our bar. The length of the bar is 1 pixel for each degree, so the *column* argument is once again the same as the value of the current array element. For example, the value of `temperatures[0]` is 57, so the width of the rectangle is 57 pixels, from pixel column 0 to pixel column 56. We want our text to begin at the next column, 57, the same value as that stored in `temperatures[0]`. Thus, our call to the `text()` function can be

```
text(temperatures[count], temperatures[count],
```

The last argument needed by the `text()` function is the pixel *row* number for the text. This is the value that is already stored in our `barTop` variable. Thus, we conclude the call to the `text()` function with

```
text(temperatures[count], temperatures[count], barTop);
```

Our program is now

```
int[] temperatures = {57, 10, 32, 78, 212, 101};

size(250, 250);

int spacer = 35;
int barHeight = 10;
int barTop;

for (int count = 0; count < temperatures.length; count = count + 1)
{
  fill(255, 0, 0);// red
  barTop = (count + 1) * spacer;
  rect(0, barTop, temperatures[count], barHeight );
  fill(0, 0, 0);// black
  text(temperatures[count], temperatures[count], barTop);
}
```

When we run this program, we see that black numbers have been added to the end of each bar.

Save and close the `GraphTemps` program.

Declaring and Creating an Array with the new Operator

Select **File > New** to start a new program. Save this program as `DefaultsArray`.

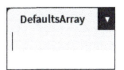

Modify this program so that it consists of the following two statements:

```
int[] temperatures = {57, 10, 32, 78, 212, 101};

printArray(temperatures);
```

Remember that in order to create an array, Processing needs to know the *size* of the array so that it can set aside sufficient space in the computer's memory for it. So far, we have been assigning values for an array in the following manner:

```
int[] temperatures = {57, 10, 32, 78, 212, 101};
```

This statement indicates to Processing how large the array needs to be in the computer's memory in order to hold these `int` values.

However, we do not always know in advance the specific values that will be stored in an array. In this case, we need to use the **new** operator, followed by the *type* and *size* of the array, in order to instruct Processing how large to make the array. For example, to declare an array named `temperatures` that can hold six values of type `int` *without* specifying what these six values are, we write the following:*

```
int[] temperatures = new int[6];
printArray(temperatures);
```

The first statement in this program causes Processing to do the following:

1) Processing sets aside locations in the computer's memory for a new array of six `int` values.

2) Processing stores in the `temperatures` variable the address in the computer's memory where this new array is located.

3) Processing initializes all the elements of this new array to 0, the default value for an `int` array.

We can visualize this behind-the-scenes process as

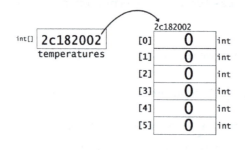

---

* We also have the option of using *two* statements here:
```
int[] temperatures;
temperatures = new int[6];
```

And we can visualize the resulting array as

temperatures	
0	0
1	0
2	0
3	0
4	0
5	0

We can verify the contents of this array by running our program and examining the resulting output:

```
[0]  0
[1]  0
[2]  0
[3]  0
[4]  0
[5]  0
```

Why did Processing initialize each element of the array to zero? Whenever we create a new array but do not specify initial values for it, Processing initializes the array with **default values**. As we've seen in our current example, 0 is the default value for an element in an array of type int. On the other hand, 0.0 is the default value for an element in an array of type float. To illustrate, let's change the type of our temperatures array to float:

```
float[] temperatures = new float[6];
printArray(temperatures);
```

Now, this first statement in our program causes Processing to do the following:

1) Processing sets aside locations in the computer's memory for a new array of six float values.

2) Processing stores in the temperatures variable the address in the computer's memory where this new array is located.

3) Processing initializes all the elements in this new array to 0.0, the default value for a float array.

We can visualize this behind-the-scenes process as

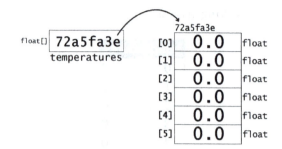

And we can visualize the resulting array as

temperatures	
0	0.0
1	0.0
2	0.0
3	0.0
4	0.0
5	0.0

When we run this modified version of our program, we now see the following output to the console:

```
[0] 0.0
[1] 0.0
[2] 0.0
[3] 0.0
[4] 0.0
[5] 0.0
```

Arrays of elements of the other data types we have studied also have default values.*

Save and close the DefaultsArray program.

Yahtzee Example: Filling an Array with Random Numbers

The game of Yahtzee uses five dice.

Let's create a program to mimic some of the aspects of this game. Select **File > New** to start a new program. Save this program as **FiveDice**.

---

* The default value of each element in an array in type char is the character with Unicode number 0, which is essentially an empty character. The default value of an array of type String is a special value, null. The default value of an array of type boolean is false.

```
FiveDice  ▼
|
```

We'll start by declaring an array to hold the `int` values of five dice. Because we do not know in advance what the values of these five dice will be, we write

```
int[] dice = new int[5];
```

This statement causes Processing to set aside space in the computer's memory for an array named `dice` to hold five values of type `int`, each of which is initialized to 0. We can visualize this behind-the-scenes process as

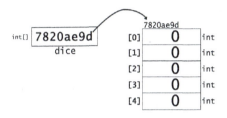

And we can visualize the resulting array as

dice	
0	0
1	0
2	0
3	0
4	0

Let's add a statement to display this array using the `printArray()` function:

```
int[] dice = new int[5];
printArray(dice);
```

When we run this program, we see the following output in the console:

```
[0] 0
[1] 0
[2] 0
[3] 0
[4] 0
```

We'll assign random values to each of the five dice. For this, we need to insert a `for` loop to count through the indexes of the array:

```
int[] dice = new int[5];
```

```
for (int count = 0; count < dice.length; count = count + 1)
{

}

printArray(dice);
```

To assign each array element a randomly chosen integer value from 1 to 6, we can insert the following statement into the body of our for loop:

```
int [] dice = new int [5];
for (int count = 0; count < dice.length; count = count + 1)
{
  dice[count] = int( random(1, 7) );
}

printArray(dice);
```

Now, when we run our program, random values from 1 to 6 are assigned as the elements of the dice array. We can verify this by examining the console output:

```
[0] 2
[1] 1
[2] 4
[3] 3
[4] 4
```

To better resemble a single roll of five dice, let's change the format of our console output:

```
int [] dice = new int [5];

for (int count = 0; count < dice.length; count = count + 1)
{
  dice[count] = int( random(1, 7) );
  print(dice[count] + " ");
}

println();
```

Now, when we run our program, these array elements are displayed on a single line in the console:

```
2 1 4 3 4
```

Processing's **min()** function can be used to find the *smaller* of two values. However, it can also be used to determine the *smallest* element in an array. For example, the following call to the min() function,

```
min(dice)
```

will return the smallest value currently found in the dice array. Similarly, Processing's **max()** function can be used to find the *larger* of two values. However, it can also be used to determine the *largest* element in an array. For example, the following call to the max() function,

```
max(dice)
```

will return the largest value currently found in the dice array.

One of the goals of the Yahtzee game is to roll a "Yahtzee," which is when the number shown on all five dice is the same.

Consider this: If the smallest and the largest value in an array are *equal*, then this means that *all* the elements in the array must all be the *same* value. Thus, if we would like to simulate rolling the dice until all five of them are the same, we can add the following do-while statement to our program:

```
int[] dice = new int[5];

do
{
  for (int count = 0; count < dice.length; count = count + 1)
  {
    dice[count] = int( random(1, 7) );
    print(dice[count] + " ");
  }
  println();
}
while ( min(dice) != max(dice) );
```

Now, when we run our program, the simulation of rolling five dice continues until all five dice are the same:

```
6 2 1 1 6
2 6 1 3 2
1 6 2 2 4
  ⋮
1 4 5 3 4
4 2 4 1 4
6 2 6 2 6
5 5 5 5 5
```

If we would also like to *count* the number of rolls it takes until all five dice are the same, we can insert the following statements:

```
int[] dice = new int[5];

int rolls = 0;

do
{
  for (int count = 0; count < dice.length; count = count + 1)
  {
    dice[count] = int( random(1, 7) );
    print(dice[count] + " ");
  }
  rolls = rolls + 1;
  println();
}
while ( min(dice) != max(dice) );

println("Number of rolls: " + rolls);
```

Now, when we run our program, we also see the total number of rolls that took place before all five dice were the same:

```
5 6 5 2 1
6 3 4 3 4
6 2 3 3 2
  ⋮
3 3 2 5 5
5 1 3 4 4
3 3 1 4 3
5 5 5 5 5
Number of rolls: 56
```

Save and close the FiveDice program.

Shorter Loop Counter Variable Names

Among experienced programmers, it is very common for the loop counter variable in a for statement to be named with a single letter, such as **i**. For example, the following for statement,

```
for (int count = 0; count < dice.length; count = count + 1)
{
  dice[count] = int( random(1, 7) );
}
```

could also be written

```
for (int i = 0; i < dice.length; i = i + 1)
{
  dice[i] = int( random(1, 7) );
}
```

However, for beginning programmers, descriptive variable names are usually the better choice, rather than single-letter variable names.

Increment (++) and Decrement (--) Operators

Most often, when we write for statements, we increment the loop counter variable by 1 each time the loop repeats. In such cases, we write the first line of such a loop as

```
for (int count = 0; count < 5; count = count + 1)
```

To *increase* the value of a variable by 1, Processing provides the **increment operator**:

```
++
```

Thus, the first line of the preceding for statement can also be written:

```
for (int count = 0; count < 5; count++)
```

Similarly, to *decrease* the value of a variable by 1, Processing provides the **decrement operator**:

```
--
```

For example, to count downward by ones in a for , loop we can write the first line of such a for statement

```
for (int count = 5; count > 0; count = count - 1)
```

Or, we can use the decrement operator and write

```
for (int count = 5; count > 0; count--)
```

Experienced programmers often favor the use of these shortcut operators. However, this may make the code less readable for less experienced programmers. Also, the use of these operators can sometimes lead less experienced programmers to forget that it is possible to increment or decrement the loop counter variables by amounts other than 1. Moreover, there are aspects of these two operators that can make them tricky when they are used outside of for statements. For all these reasons, we have not used the increment and decrement operators in this book.

Interactive Example: Inputting Names

We can also make arrays consisting of elements of the *nonnumeric* types we've studied: boolean, char, and String.

For example, suppose we would like to input four student names into an array. Select **File > New** to start a new program. Save this program as **StringArray**.

We'll start with the import statement we need to use dialogs:

```
import static javax.swing.JOptionPane.*;
```

We need to declare an array of type String to hold the student names:

```
String[] students = new String[4];
```

Next, we'll use a for statement to repeatedly input a name for each element of the array:

```
for (int count = 0; count < students.length; count = count + 1)
{
   students[count] = showInputDialog("Enter name of student:");
}
```

Finally, let's display the list of names once they are read into the array:

```
printArray(students);
```

Our program is now

```
import static javax.swing.JOptionPane.*;

String[] students = new String[4];

for (int count = 0; count < students.length; count = count + 1)
{
   students[count] = showInputDialog("Enter name of student:");
}

printArray(students);
```

When we run this program, we are prompted four times to enter a student name, each of which is entered into the array:

We can visualize the resulting array as

	students
0	"Rebecca"
1	"Megan"
2	"Sara"
3	"Callie"

Afterward, the names in the array are displayed to the console:

```
[0]  "Rebecca"
[1]  "Megan"
[2]  "Sara"
[3]  "Callie"
```

One of the difficulties of working with an array is that because of its fixed size, we need to know in advance how many values will be stored so that we can make the array large enough. In the current example, if we would like the user to specify the number of students that will be entered, we can modify our program so that it is the following:

```
import static javax.swing.JOptionPane.*;

String input = showInputDialog("Number of students:");
int size;
size = int(input);

String[] students = new String[size];

for (int count = 0; count < students.length; count = count + 1)
{
  students[count] = showInputDialog("Enter name of student:");
}

printArray(students);
```

Now, when we run our program, the user is asked how many students will be entered:

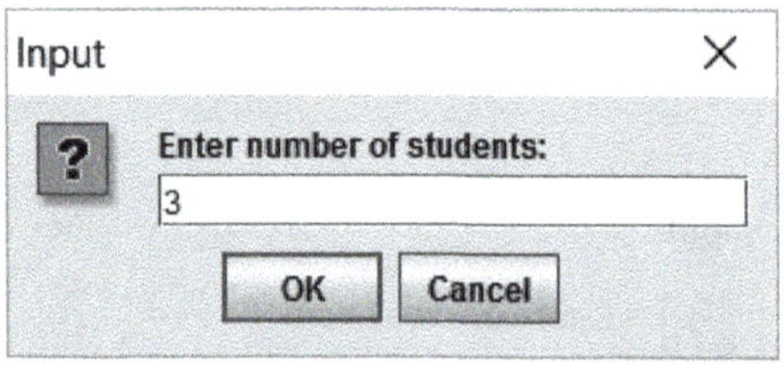

An array of the specified size is then created. Afterward, the user is repeatedly prompted to enter a student name, as many times as the number of students specified:

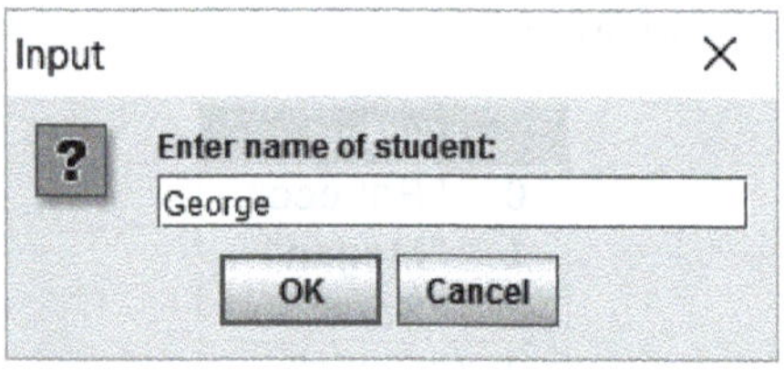

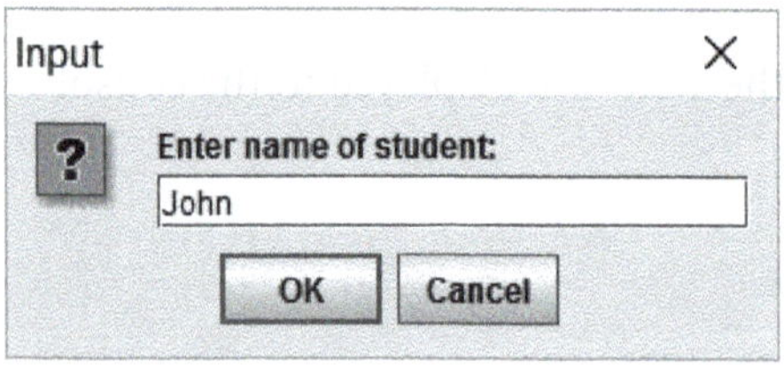

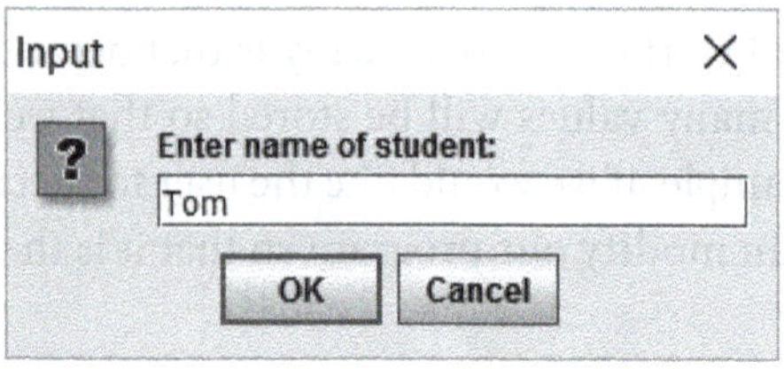

The array created as a result of these entries can be visualized as

	students
0	"George"
1	"John"
2	"Tom"

We can see the contents of this array represented in the console output produced by the call to the printArray() function:

```
[0] "George"
[1] "John"
[2] "Tom"
```

Save and close the StringArray program.

Searching an Array: Linear Search

Another common operation performed on an array is *searching* for a particular value. One of the least efficient but simplest methods of searching is the **linear search**, where we simply step through the array elements in order until we find the value we are seeking or until we reach the end of the array.

Let's construct a program to do a basic linear search. Select **File > New** to start a new program. Save this program as **LinearSearch**.

```
LinearSearch   ▼
|
```

We'll once again use our temperatures array:

```
int[] temperatures = {57, 10, 32, 78, 212, 101};
```

Let's search this array for a certain value. We'll declare and initialize a variable to hold the particular value we are seeking:

```
int[] temperatures = {57, 10, 32, 78, 212, 101};

int seeking = 78;
```

Let's also display this value to the console:

```
int[] temperatures = {57, 10, 32, 78, 212, 101};

int seeking = 78;

println("Searching for: " + seeking);
```

Next, we'll need a `for` statement to count through the indexes of the array:

```
int[] temperatures = {57, 10, 32, 78, 212, 101};

int seeking = 78;

println("Searching for: " + seeking);

for (int count = 0; count < temperatures.length; count = count + 1)
{
}
```

As we count through this `for` loop, let's display each element of the array:

```
int[] temperatures = {57, 10, 32, 78, 212, 101};

int seeking = 78;

println("Searching for: " + seeking);

for (int count = 0; count < temperatures.length; count = count + 1)
{
   println("temperatures[" + count + "] = " + temperatures[count] );
}
```

When we run this program, we see the following output to the console:

```
Searching for: 78
temperatures[0] = 57
temperatures[1] = 10
temperatures[2] = 32
temperatures[3] = 78
temperatures[4] = 212
temperatures[5] = 101
```

Next, at each step of the `for` loop, we wish to check if the current array element is equal to the value we are seeking. If the current array element is equal to the value we are seeking, we announce this in the console. To achieve this, we can insert the following `if` statement inside the body of the `for` statement:

```
int[] temperatures = {57, 10, 32, 78, 212, 101};

int seeking = 78;

println("Searching for: " + seeking);
```

```
for (int count = 0; count < temperatures.length; count = count + 1)
{
  println("temperatures[" + count + "] = " + temperatures[count] );

  if (temperatures[count] == seeking)
  {
    println(seeking + " has been found at [" + count + "]");
  }
}
```

Also, after our `for` statement, we'll display a message in the console announcing that the search of the array is finished.

```
int[] temperatures = {57, 10, 32, 78, 212, 101};

int seeking = 78;

println("Searching for: " + seeking);

for (int count = 0; count < temperatures.length; count = count + 1)
{
  println("temperatures[" + count + "] = " + temperatures[count] );

  if (temperatures[count] == seeking)
  {
    println(seeking + " has been found at [" + count + "]");
  }
}

println("end of search");
```

When we run this program, we see the following output to the console:

```
Searching for: 78
temperatures[0] = 57
temperatures[1] = 10
temperatures[2] = 32
temperatures[3] = 78
78 has been found at [3]
temperatures[4] = 212
temperatures[5] = 101
end of search
```

If we would like the user to be able to specify the temperature to seek in the array, we can change our program to the following:

```
import static javax.swing.JOptionPane.*;

int[] temperatures = {57, 10, 32, 78, 212, 101};

String input;
input = showInputDialog("Enter temperature for search:");

int seeking;
seeking = int(input);

println("Searching for: " + seeking);

for (int count = 0; count < temperatures.length; count = count + 1)
{
  println("temperatures[" + count + "] = " + temperatures[count] );

  if (temperatures[count] == seeking)
  {
    println(seeking + " has been found at [" + count + "]");
  }
}

println("end of search");
```

Now, when we run our program, the user is prompted to enter a temperature,

and this entered temperature is used for the search of the array, as we see in the console output:

```
Searching for: 212
temperatures[0] = 57
temperatures[1] = 10
temperatures[2] = 32
temperatures[3] = 78
temperatures[4] = 212
212 has been found at [4]
temperatures[5] = 101
end of search
```

Save and close the LinearSearch program.

Summary

In this chapter, we learned about an important data structure known as an **array**.

- An array consists of a *fixed* number of items of the *same type* arranged in a *specific order*.

- An item in an array is known as an **element**.

- We can declare and initialize an array using *specific values*.

- We refer to an element of an array using an **index**, also known as a **subscript**.

- Array indexes are zero based; in other words, the first item in the array has 0 as its index.

- We can declare and create an array with the **new** operator; the elements of such an array will be initialized to a *default* value that is appropriate to the type of an array.

- An array variable actually stores the *address* where the array is located in the computer's memory.

- Using an index that is greater than or equal to the length of the array can cause an error known as an **out-of-bounds exception**.

- Counting loops, especially `for` loops, are useful for processing arrays.

- An array is a kind of object, evidenced by its **length** property, which stores the length, of a given array.

- Processing provides several functions that can be used with arrays, including **printArray()**, **min()**, and **max()**.

Processing also provides a number of other functions for sorting and copying an array. Arrays can also be passed as arguments to a function. To learn about such uses of arrays, see the online chapter "Advanced Array Operations."

Processing also has the capability for *multi*dimensional arrays. To learn about these, see the online chapter "Two-Dimensional Arrays."

Exercises

1) Write a program to do the following:

   a) Declare an array to hold seven `int` values.

   b) Display the default elements in the console.

2) Write a program to do the following:

   a) Declare an array to hold 15 `float` values.

   b) Display the default elements in the console.

3) Create an array that will store the five vowel characters. Then, display these elements to the console. The output should be

```
[0] 'a'
[1] 'e'
[2] 'i'
[3] 'o'
[4] 'u'
```

4) Create an array that will store the names of all the days of the week in it. Then, display these elements to the console. The output should be

```
[0] "Sunday"
[1] "Monday"
[2] "Tuesday"
[3] "Wednesday"
[4] "Thursday"
[5] "Friday"
[6] "Saturday"
```

5) Write a program to declare and initialize an array to contain the following values:

0

15

20

35

45

50

70

Then, display these array elements to the console.

6) Revise the program in the preceding exercise to use a `for` statement to output the values in the following format:

```
element 0 is 0
element 1 is 15
element 2 is 20
element 3 is 35
element 4 is 45
element 5 is 50
element 6 is 70
```

7) Revise the program in the previous exercise so that the sum of the elements is also calculated and displayed.

8) Revise the program in the previous exercise so that the average of the elements is also calculated and displayed.

9) Write a program to do the following:

a) Declare an array to hold 10 integers.

b) Fill the array with random integers between 1 and 100.

c) Display the array elements in the console.

10) Revise the program in the previous exercise so that the `max()` and `min()` functions are used to find largest and smallest elements, and these are also displayed.

11) Write a program to do the following:

a) Set the canvas size to 150 pixels by 200 pixels.

b) Declare an array named `columns` to hold five integers.

c) Declare an array named `rows` to hold five integers.

d) Fill each element of the `columns` array with a random integer from 0 to the width of the canvas.

e) Fill each element of the `rows` array with a random integer from 0 to the height of the canvas.

f) Display the value of each corresponding row and column on the same line in the console.

g) Using each element in the `columns` array as a column number of the center of a circle and each element in the `rows` array as the corresponding row number of the center of the same circle, draw a circle with a diameter of 20 at that row and column location.

Sample console output

```
column: 74, row: 88
column: 23, row: 33
column: 107, row: 83
column: 115, row: 24
column: 53, row: 146
```

Sample canvas output

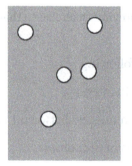

CHAPTER 10

Introduction to Objects

In processing, we have learned to work with *primitive* types of data, such as `int`, `float`, `char`, and `boolean`. However, we have also learned that Processing allows us to work with data *objects*, such as a `String` object or an array. In this chapter, we will undertake a basic introduction to *creating* our own *type* of objects by creating a new *class*.

Understanding an Object

What exactly is an *object*, and what makes it special? In Processing, an **object** can bundle together

1) **Attributes**—*variables* that store items of information about the object;*

2) **Methods**—*functions* that enable the object to perform certain actions involving the object's attributes (variables).†

Thus, whenever we have items of *information* that are related to each other and that we regularly use to perform certain actions, we might define a new type of object to bundle this information together with this *usage*.

One very common reason for defining a new type of Processing object is to model (represent) a certain thing that we use in life. For example, a stopwatch involves *information* (e.g., the elapsed time) and *usage* through actions involving its information (e.g., pressing the reset button resets the elapsed time).

---

* An attribute is also known by a variety of other terms, such as *field*, *state*, *characteristic*, and *data member*.
† A method is simply a function that is defined inside a class.

Defining a New Type of Object By Creating a New Class

Select **File** > **New** to start a new program. Save this program as `ObjectsAndClasses`.

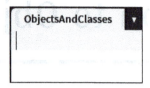

The general definition of a particular type of object is known as a **class**. For example, behind the scenes, Processing's `String` *class* definition provides the general definition of what a `String` *object* is. This class enables us to declare variables of type `String`.

We can also create a *new* type of object by creating a new *class* of objects. In such a class definition, we define

1) *Variables* that will store items of information that we associate with this type of object

2) *Functions* that make use of the attribute variables to provide us with the capability for certain actions that we associate with this type of object

For example, suppose we wish to create a new kind of object called a `Circle`. A circle on the Processing canvas has certain items of *information* associated with it, for example, the pixel column and pixel row location of its center. We can also *use* this object through certain *actions* that involve its items of information, for example, drawing the circle on the canvas using the pixel column and pixel row. We can create a new `Circle` type of object by defining a new `Circle` class. To do this, we will define

1) *Variables* to store the row and column location of the circle's center, along with other attributes we associate with a circle.

2) *Functions* that allow us to use this circle through performing certain actions involving its attribute variables.

We need to be in *active mode* if we wish to define and use a new class. Let's start with a `setup()` function that will create a 200-pixel by 200-pixel canvas:

```
void setup()
{
  size(200, 200);
}
```

We can now add a class definition to our program.*

---

* This would often be done using a separate file or a separate tab in Processing. However, to keep our code more compact, we will insert our new class definition into the current program file.

A class definition begins with the keyword `class`, followed by the *name* of the type of object we are creating. The first letter of a class name should be capitalized.* Thus, we begin our definition of a `Circle` class by adding the following statement to our program *after* the `setup()` function:

```
class Circle
```

The *body* of the class definition that follows must be enclosed inside a pair of curly braces. Thus, our program is now

```
void setup()
{
  size(200, 200);
}

class Circle
{

}
```

Next, we'll define the *variables* for the class *attributes*, the items of information about this new kind of object that we wish to store. Each `Circle` object should have

- A variable to store the pixel *column* of its center

- A variable to store the pixel *row* of its center

- A variable to store its *diameter*

Thus, inside the class body, we declare three variables for these attributes:

```
class Circle
{
  int column;
  int row;
  int diameter;
}
```

As a result, *every* object of type `Circle` will now have these three attribute variables: a column, a row, and a diameter. **The *scope* of any attribute variable in a class is the *entire* class.** Thus, these three variables may be used throughout the entire `Circle` class definition.

---

* This is why `String` is always capitalized.

Next, let's define a *function* in this class of object that will enable us to perform some *action* involving one or more of the attribute variables. A function defined inside a class is also known as a **method**.

Every Circle object should have a method (function) to draw the circle on the canvas using its column, row, and diameter attribute variables. We'll define a method named render() to perform this drawing. This render() method does not need to return a value, so we will declare its return type to be void. Also, no parameters are necessary for the render() method (function), because the three internal attribute variables we've defined—column, row, and diameter—provide all the information that is needed to draw the circle using a call to the ellipse() function, and these three variables can be used throughout the entire Circle class. Thus, inside the class, we write the definition of the render() method as follows:

```
class Circle
{
  int column;
  int row;
  int diameter;

  void render()
  {
    ellipse(column, row, diameter, diameter);
  }

}
```

Given our current Circle class definition, every object of type Circle will have a built-in method (function) named render(), which, when called, will cause that Circle object to be rendered on the canvas.

When working with objects, it is common practice to adopt a kind of perspective that treats each object as if it is a kind of "character" that has a mind of its own. From this perspective, instead of saying that we want to "call the object's render() method," we would say that we want to "ask the object to render itself" or to "ask the object to perform its render() method."

We now have a Circle class that contains a general definition of what a Circle object is. Thus, we might say that the Circle class provides a kind of *blueprint* for any Circle object that we wish to create.

Save the `ObjectsAndClasses` program. We will continue working with it in the next section.

Instantiating an Object

Return to the `ObjectsAndClasses` program. Once a new *class* of object has been defined, this gives us a definition of a new *type* of object. Thus, we can use this class to create objects of the new type that this class defines.

In terms of our blueprint analogy, the `Circle` class provides a kind of "blueprint" for what a `Circle` object is in our Processing program. Thus, we can use this `Circle` class blueprint as the basis for creating an actual `Circle` object. For example, let's create a `Circle` object named `littleCircle` that has 30 as its `column`, 50 as its `row`, and 20 as its `diameter`.

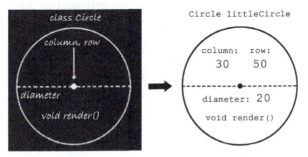

How do we create an actual `Circle` object? As usual, we start with the declaration of a variable. Because of our `Circle` class, we can now declare a variable named `littleCircle` that is of type `Circle`:

```
void setup()
{
  size(200, 200);

  Circle littleCircle;

}
```

As we learned in our study of arrays and `String` objects, a variable that is a type of object does not contain an actual value of that type; rather, it contains the *address* in the computer's memory where the object is actually stored. Similarly, when we first declare the `littleCircle` variable, this variable that is designed to hold the address of an object does not yet contain the address of an object. We can visualize this as

We also have learned that a statement declaring an array does not actually create a new array; rather, we must use the `new` operator in order to create a new array with default values. Similarly, to perform the action of actually creating a `Circle` object, we need to use the **new** operator:

```
void setup()
{
  size(200, 200);

  Circle littleCircle;
  littleCircle = new Circle();

}
```

The new operator in our added statement causes Processing to set aside the necessary space in memory to hold the contents of a Circle object and store the address of this memory location in the littleCircle variable. Also, as in the case of an array, each of the object's attribute variables of primitive type is initialized to the *default* value for that type. Thus, in this particular case, the three int attribute variables of the littleCircle object are all initialized to 0. We can visualize this behind-the-scenes result as

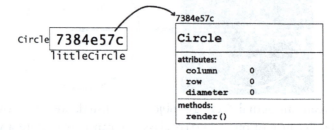

Inserting the following statement into our setup() function, will enable us to verify that our littleCircle variable now contains a memory address:

```
print(littleCircle);
```

Thus, our program is now

```
void setup()
{
  size(200, 200);

  Circle littleCircle;
  littleCircle = new Circle();

  print(littleCircle);

}

class Circle
{
  int column;
  int row;
  int diameter;
```

```
  void render()
  {
    ellipse(column, row, diameter, diameter);
  }
}
```

When we run this program, our console output is now something like

ObjectsAndClasses$Circle@7384e57c

Again, the characters after the @ symbol are hexadecimal digits describing a memory location.

In object-oriented programming (OOP) terminology, we have now created an **instance** of the Circle class. Similarly, this generation of an individual *object* of a type that is defined in a *class* is known as **instantiating** an object. Thus, we can say that we have now successfully created (instantiated) our first object (instance) of type Circle.

We can now delete the statement containing the call to the print() function.

~~print(littleCircle);~~

Save the ObjectsAndClasses program. We will continue working with it in the next section.

Setting Attributes

Return to the ObjectsAndClasses program. Like any object of type Circle, our littleCircle object has its *own* set of the three internal variables that represent its three attributes: column, row, and diameter. Such internal attribute variables that belong to each individual object are sometimes known as **instance variables**.

By default, each of these variables is initialized to the value 0, because they are of type int. Let's now change these attributes by assigning new values to these variables. For now, we'll use *dot notation* to do so. We can refer to the attribute variable of an object by writing the *object* variable name, followed by a *dot* (period) and then the *attribute* variable name:

```
object.attribute
```

Thus, in this particular case, we can assign values to our Circle object's attribute variables by writing

```
littleCircle.column = 30;
littleCircle.row = 50;
littleCircle.diameter = 20;
```

After inserting these statements, our setup() function is now

```
void setup()
{
  size(200, 200);

  Circle littleCircle;
  littleCircle = new Circle();

  littleCircle.column = 30;
  littleCircle.row = 50;
  littleCircle.diameter = 20;
}
```

We can visualize the behind-the-scenes result of these three statements as

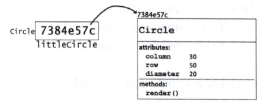

Save the ObjectsAndClasses program. We will continue working with it in the next section.

Using Methods

Return to the ObjectsAndClasses program. Now that the three attributes have been set to the values we desire, we are ready to ask the littleCircle object to draw itself on the canvas using the render() method (function) that is built into each Circle object. This render() method is an example of what is known as an **instance method**. We can also use dot notation to create a call to an object's instance method:

```
littleCircle.render();
```

Thus, our complete program is now

```
void setup()
{
  size(200, 200);

  Circle littleCircle;
  littleCircle = new Circle();

  littleCircle.column = 30;
  littleCircle.row = 50;
  littleCircle.diameter = 20;

  littleCircle.render();
}
```

```
class Circle
{
  int column;
  int row;
  int diameter;

  void render()
  {
    ellipse(column, row, diameter, diameter);
  }
}
```

When we run this program, we now see the following output to the canvas:

Thus, the littleCircle object of type Circle has been successfully created (instantiated), its attribute instance variables have been initialized, and it has successfully drawn itself on the canvas in response to our call to its render() instance method.

Save the ObjectsAndClasses program. We will continue working with it in the next section.

Defining a Constructor

Return to the ObjectsAndClasses program. As we have learned, our statement containing the new operator is necessary for the actual creation (instantiation) of the individual littleCircle object of type Circle. Let's take a closer look at this statement. Notice that the portion of this statement following the new operator looks much like a function call:

```
littleCircle = new Circle();
```

Indeed, something very much like a function (method) call is taking place here. Specifically, this is a call made by the new operator to the Circle() **constructor**. When the new operation is performed in our statement, Processing calls the Circle() constructor. As a result, memory is set aside for the littleCircle object and its instance variables are initialized to their default values.

Where is this Circle() constructor defined? The Circle() constructor that is currently being called when we run our program is the **default constructor**. Processing automatically generates this default constructor for us if we have not defined a constructor ourselves. However, it is actually better for us to provide an explicit definition of a constructor ourselves.

Directly accessing an object's attribute *variables* from *outside* the class definition using dot notation violates a fundamental principle of OOP known as **encapsulation**. According to this principle, we really should consider an object's attribute variables to be "private" variables that are encapsulated *inside* the object and should not be manipulated by any statements *outside* the object. Instead, we should *initialize* attribute instance variables by using a constructor. Similarly, we should only *change* or *retrieve* the values of attribute instance variables by using an instance method designed for this purpose. Thus, according to the principle of encapsulation, we should *not* be manipulating the attribute instance variables encapsulated inside the `littleCircle` object. However, this is exactly what we are doing by means of the following statements that are currently in the `setup()` function

```
littleCircle.column = 30;
littleCircle.row = 50;
littleCircle.diameter = 20;
```

Instead, we should define and use our own constructor for the `Circle` class. We can then use this constructor to initialize the instance variables of a `Circle` object to whatever specific values we choose, instead of the default values that were assigned by the default `Circle()` constructor.

Defining the *header* of a constructor is similar to defining the header of a function, except a *return type* is *not* specified. Not even the keyword `void` should appear in the constructor's header. Thus, we begin the header with simply the *name* of the constructor, which is always the *same* as the name of the class. Also, unlike functions, the *first letter* of a constructor's name is always *capitalized*. A constructor is typically defined *after* any instance variables that are declared in the class and *before* any instance methods that will be defined in the class. Thus, we can start our constructor definition by inserting the following into our class:

```
class Circle
{
  int column;
  int row;
  int diameter;

  Circle

  void render()
  {
    ellipse(column, row, diameter, diameter);
  }
}
```

Now, let's declare the parameters for this constructor. We want to be able to use this constructor to specify initial values to assign to each of the three attribute instance variables of the class: `column`, `row`, and `diameter`. Thus, we need to define three *parameter* variables, one corresponding to each of these three attribute instance variables. As in a function, we place the declarations of these parameter variables inside the header of our constructor:

```
Circle(int columnInitial, int rowInitial, int diameterInitial)
```

Thus, the *header* of a constructor is much like that of a function (method), but it is also different in several ways:

- The constructor for a class always has the *same name* as the class itself.

- The first letter in the name of a constructor is *capitalized*.

- Although a constructor does not return a value, we do not use the keyword `void` in the definition of a constructor.

The header of our new constructor is complete.

We can now define the *body* of the constructor. This is where we initialize the attribute variables to the desired values. In this particular case, we will use our three parameters to initialize each of the three instance variables:

```
Circle(int columnInitial, int rowInitial, int diameterInitial)
{
  column = columnInitial;
  row = rowInitial;
  diameter = diameterInitial;
}
```

Our class definition is now

```
class Circle
{
  int column;
  int row;
  int diameter;

  Circle(int columnInitial, int rowInitial, int diameterInitial)
  {
    column = columnInitial;
    row = rowInitial;
    diameter = diameterInitial;
  }
```

```
    void render()
    {
      ellipse(column, row, diameter, diameter);
    }
  }
```

Now, let's return to the setup() function. We see that underlining has appeared in the following statement:

```
littleCircle = new Circle();
```

The corresponding error message is

The constructor "Circle()" does not exist

This message has appeared because, once we declare a constructor ourselves, Processing no longer creates the default Circle() constructor with no parameters that we are currently invoking in this statement. Instead, we need to use the constructor that we have defined and expects three arguments, one for each of the three parameter variables defined in the header of the constructor:

```
Circle(int columnInitial, int rowInitial, int diameterInitial)
```

Thus, we need to supply three int arguments in the call to the Circle() constructor for the littleCircle object's column, row, and diameter:

```
littleCircle = new Circle(30, 50, 20);
```

As a result, we now have a match between the argument list in the call to the constructor and the parameter list in the header of the constructor method. Thus, the underlining and error message disappear.

Our new constructor is defined to do the work of initializing the object's attribute instance variables:

```
Circle(int columnInitial, int rowInitial, int diameterInitial)
{
  column = columnInitial;
  row = rowInitial;
  diameter = diameterInitial;
}
```

Thus, we no longer need to manipulate the attribute variables from outside of our littleCircle object using dot notation. Let's delete these three statements in the setup() function that violated the principle of encapsulation:

```
littleCircle.column = 30;
littleCircle.row = 50;
littleCircle.diameter = 20;
```

Thus, our program is now

```
void setup()
{
  size(200, 200);

  Circle littleCircle;
  littleCircle = new Circle(30, 50, 20);

  littleCircle.render();

}

class Circle
{
  int column;
  int row;
  int diameter;

  Circle(int columnInitial, int rowInitial, int diameterInitial)
  {
    column = columnInitial;
    row = rowInitial;
    diameter = diameterInitial;
  }

  void render()
  {
    ellipse(column, row, diameter, diameter);
  }
}
```

When we run this program, the littleCircle object has been created (instantiated) by means of its constructor, and its attribute instance variables are initialized to the values that we specified in the constructor call. We can visualize this result as

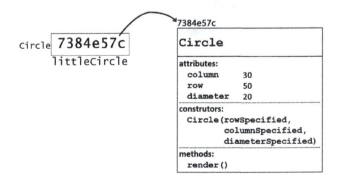

The littleCircle object has also drawn itself on the canvas by means of its render() method:

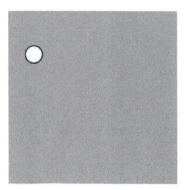

Save the `ObjectsAndClasses` program. We will continue working with it in the next section.

Multiple Objects

Return to the `ObjectsAndClasses` program. One of the benefits of defining a new class is that it enables us to create *multiple* objects of the new type that the class defines. For example, let's create a *second* object of type `Circle` named `bigCircle`. We will then have two objects created from the "blueprint" that our `Circle` class provides. We can visualize this as

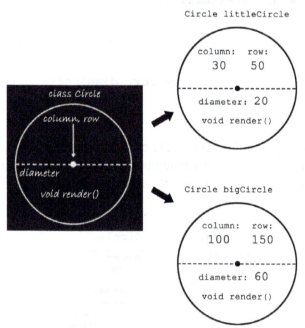

We'll start with a variable declaration of our new second object:

```
Circle bigCircle;
```

Next, we instantiate this object using the new operator and a call to our constructor:

```
bigCircle = new Circle(100, 150, 60);
```

Let's add a call to have this second object render itself as well:

```
bigCircle.render();
```

Thus, our program is now

```
void setup()
{
  size(200, 200);

  Circle littleCircle;
  littleCircle = new Circle(30, 50, 20);

  Circle bigCircle;
  bigCircle = new Circle(100, 150, 60);

  littleCircle.render();
  bigCircle.render();
}

class Circle
{
  int column;
  int row;
  int diameter;

  Circle(int columnInitial, int rowInitial, int diameterInitial)
  {
    column = columnInitial;
    row = rowInitial;
    diameter = diameterInitial;
  }

  void render()
  {
    ellipse(column, row, diameter, diameter);
  }
}
```

Now, when we run our program, *two* objects are instantiated, which we can visualize as

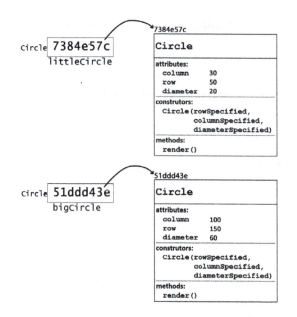

Also, both the `littleCircle` object and the `bigCircle` object render themselves on the canvas when the `render()` method of each object is called:

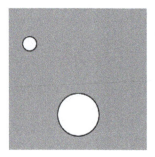

We now have created (instantiated) two objects of type `Circle`. Each of these two objects has its *own* set of the three instance variables representing its attributes: `column`, `width`, and `diameter`. Also, as we have seen, the `render()` instance method can be called *independently* for each object (instance).

Save the `ObjectsAndClasses` program. We will continue working with it in the next section.

Interacting with Objects

Return to the `ObjectsAndClasses` program. Let's make our `Circle` objects *interactive* so that we can drag them around the canvas. For this, we'll need to add a `draw()` function to our program. Recall that as soon as the `setup()` function is complete, Processing *automatically* calls the `draw()` function *repeatedly*, over and over again, until we stop our program. We'll want to repeatedly redraw the entire canvas in order to create the illusion that we're dragging our circles around the canvas. Thus, inside our `draw()` function, let's start by inserting a call to the `background()` function that will set all the pixels of the canvas to the default gray color:

```
void draw()
{
  background(204, 204, 204);
}
```

Let's also relocate the two statements that call the `render` method so that they are instead in the `draw()` function:

```
void draw()
{
  background(204, 204, 204);

  littleCircle.render();
  bigCircle.render();
}
```

As soon as we do this, underlining appears in these two statements:

```
littleCircle.render();
bigCircle.render();
```

The error messages corresponding to this underlining are

The name "littleCircle" cannot be recognized

and

The name "bigCircle" cannot be recognized

These errors are appearing because `littleCircle` and `bigCircle` are currently *local* variables. They cannot be used outside the `setup()` function. However, if we move the two declaration statements *outside* the `setup()` function, then the two variables become **global** variables that can be used in *both* the `setup()` and `draw()` functions. After this change, our program is now

```
Circle littleCircle;
Circle bigCircle;

void setup()
{
  size(200, 200);

  littleCircle = new Circle(30, 50, 20);

  bigCircle = new Circle(100, 150, 60);
}
```

```
void draw()
{
  background(204, 204, 204);

  littleCircle.render();
  bigCircle.render();
}

  class Circle
  {
    int column;
    int row;
    int diameter;

  Circle(int columnInitial, int rowInitial, int diameterInitial)
  {
    column = columnInitial;
    row = rowInitial;
    diameter = diameterInitial;
  }

  void render()
  {
    ellipse(column, row, diameter, diameter);
  }
}
```

Now, we'll create another instance method inside our Circle class definition that will enable each Circle object to update its position. We do not need this method to return a value, so it will be a void method.

```
void update()
{

}
```

We'll have this update() method check whether or not the mouse pointer is currently over the rendering of this Circle object on the canvas by determining whether or not the distance between the center of the circle and the tip of the mouse pointer is less than the radius of the circle. We'll declare some local variables for this. First, we'll declare a variable to store the radius of the Circle object:

```
void update()
{
  int radius = diameter / 2;
}
```

Next, we'll declare a variable to store the distance between the tip of the mouse pointer and the center of the `Circle` object. To calculate this distance, we can use Processing's **dist()** function, which measures the pixel distance between two pixel locations on the canvas. The general form of this function is

```
dist(row₁, col₁, row₂, col₂)
```

The dist() function returns the distance between the two pixel locations as a **float** value. We'll assign this calculated distance to a float variable mouseDistance:

```
void update()
{
  int radius = diameter / 2;
  float mouseDistance = dist(mouseX, mouseY, column, row);
}
```

Both radius and mouseDistance are *local* variables: they cannot be used outside the update() method. However, as we know, the attribute instance variables—column, row, and diameter—can be used *throughout* the definition of the Circle class.

We need to determine whether or not this calculated distance between the mouse pointer and the pixel location is less than the radius of the circle. If so, then we know the mouse pointer is over the circle rendered on the canvas for this object. Thus, our if statement begins with

```
if (mouseDistance < radius)
```

If the mouse pointer is over the circle on the canvas, then let's update the pixel location of the Circle object to the current pixel location of the mouse pointer:

```
if (mouseDistance < radius)
{
  column = mouseX;
  row = mouseY;
}
```

Thus, our Circle class definition is now

```
class Circle
{
  int column;
  int row;
  int diameter;
```

```
Circle(int columnInitial, int rowInitial, int diameterInitial)
{
  column = columnInitial;
  row = rowInitial;
  diameter = diameterInitial;
}

void render()
{
  ellipse(column, row, diameter, diameter);
}

void update()
{
  int radius = diameter / 2;
  float mouseDistance = dist(mouseX, mouseY, column, row);

  if (mouseDistance < radius)
  {
    column = mouseX;
    row = mouseY;
  }
}
}
```

As a result, each `Circle` object will now have the ability to update its position whenever we call its `update()` method using dot notation:

```
littleCircle.update();
```

or

```
bigCircle.update();
```

When the `update()` method is called and the mouse pointer is over the `Circle` object, the pixel location of the `Circle` object's center will be changed to the pixel location of the mouse pointer. However, we only want to call this new `update()` method of the `Circle` object when the mouse is pressed. Fortunately, Processing has a built-in `boolean` variable named **mousePressed** that is `true` whenever the mouse button is pressed and `false` otherwise. To call the `update()` methods of our two `Circle` objects whenever the mouse button is pressed, we add the following to the `draw()` function:

```
if (mousePressed)
{
  littleCircle.update();
  bigCircle.update();
}
```

Thus, our program is now

```
Circle littleCircle;
Circle bigCircle;

void setup()
{
  size(200, 200);

  littleCircle = new Circle(30, 50, 20);
  bigCircle = new Circle(100, 150, 60);
}

void draw()
{
  background(200, 200, 200);
  littleCircle.render();
  bigCircle.render();

  if (mousePressed)
  {
    littleCircle.update();
    bigCircle.update();
  }
}

class Circle
{
  int column;
  int row;
  int diameter;

  Circle(int columnInitial, int rowInitial, int diameterInitial)
  {
    column = columnInitial;
    row = rowInitial;
    diameter = diameterInitial;
  }

  void render()
  {
    ellipse(column, row, diameter, diameter);
  }

  void update()
  {
    int radius = diameter / 2;
    float mouseDistance = dist(mouseX, mouseY, column, row);
```

```
    if (mouseDistance < radius)
    {
      column = mouseX;
      row = mouseY;
    }
  }
}
```

When we run our program, two `Circle` objects are instantiated behind the scenes, which we can now visualize as

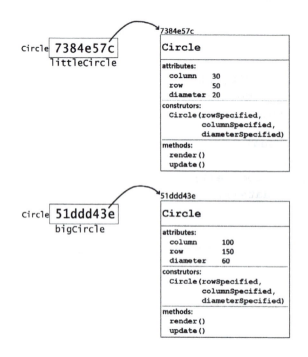

Also, because of the `update()` method of each object that is called when the mouse is pressed, we are now able to click and drag each of the circles around the canvas:

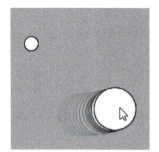

Save the `ObjectsAndClasses` program. We will continue working with it in the next section.

Arrays of Objects

Return to the ObjectsAndClasses program. In creating our Circle class, we have created a new kind of object. Thus, we can declare any number of objects of type Circle that we wish. We can also have *arrays* of objects. To illustrate, let's create an array of Circle objects.

Select **File > Save As** and resave the ObjectsAndClasses program as **ObjectArray**.

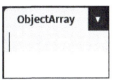

We can create an array of Circle objects by modifying our main program to the following:

```
int numberOfCircles = 5;
Circle[] circles;

void setup()
{
  size(200, 200);

  int startColumn;
  int startRow;
  int startDiameter;

  circles = new Circle[numberOfCircles];
  for (int count = 0; count < circles.length; count = count + 1)
  {
    startColumn = int(random(width));
    startRow = int(random(height));
    startDiameter = int(random(20, 60));
    circles[count] = new Circle(startColumn, startRow, startDiameter);
  }
}

void draw()
{
  background(200, 200, 200);

  for (int count = 0; count < circles.length; count = count + 1)
  {
    circles[count].render();
  }

  if (mousePressed)
  {
   for (int count = 0; count < circles.length; count = count + 1)
   {
```

```
      circles[count].update();
    }
  }
}

class Circle
{
  int column;
  int row;
  int diameter;

  Circle(int columnInitial, int rowInitial, int diameterInitial)
  {
    column = columnInitial;
    row = rowInitial;
    diameter = diameterInitial;
  }

  void render()
  {
    ellipse(column, row, diameter, diameter);
  }

  void update()
  {
    int radius = diameter / 2;
    float mouseDistance = dist(mouseX, mouseY, column, row);

    if (mouseDistance < radius)
    {
      column = mouseX;
      row = mouseY;
    }
  }
}
```

We do not need to make any changes to our Circle class. We can visualize the behind-the-scenes creation of this array of Circle objects as

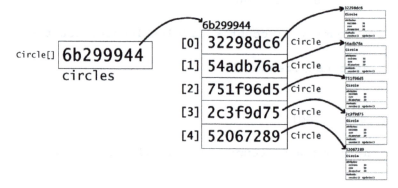

Now, when we run our program, we see five `Circle` objects that can each be dragged individually around the canvas:

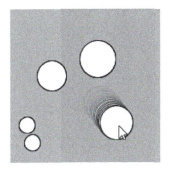

Save and close the `ObjectsAndClasses` program.

Object-Oriented Programming (OOP)

At the outset of this book, we built our programs out of simple statements. This is sometimes called **imperative programming**. Eventually, we arranged these statements into functions, marking a move into what is known as **procedural programming**.

Now that we have had an introduction to objects, we have taken steps in the direction of **object-oriented programming**, also known as **OOP**. In OOP languages such as **Java**, the fundamental unit for building programs is an object, and accordingly, every program that is written is comprised of one or more classes. Moving on to OOP is a challenging step that is beyond the scope of this book. However, after having learned what you have through this book, you are now in an excellent position to begin to make this step if you choose to do so. Congratulations!

Summary

This chapter provided a very brief introduction to creating our own type of objects by creating a new class.

- An **object** is a special type of structure that contains

  1) Variables that serve as *attributes* of the object

  2) *Functions* known as *methods*, which make use of the attribute variables.

- A **class** defines a type of object.

- An object is also known as an **instance** of a particular class.

- A **constructor** has similarities to a function (method) and is used to create an object in memory and to initialize the attribute variables of an object.

- A constructor is called using the **new** operator.

- Using a constructor to create a new object is also known as **instantiating** that object.

- Multiple objects of the same class can be instantiated within a single program.

- Arrays of objects can be created.

Exercises

1) Type in the following program and run it:

```
void setup()
{
  size(200, 200);

  Square firstSquare = new Square();
  firstSquare.render();

  Square secondSquare = new Square();
  secondSquare.render();
}

class Square
{
  int column;
  int row;
  int side;

  Square()
  {
    side = int( random(50, 100));
    column = int( random(width - side));
    row = int( random(height - side));
  }

  void render()
  {
    rect(column, row, side, side);
  }
}
```

2) Type in the following program and run it:

```
void setup()
{
  size(400, 400);

  Target bigTarget = new Target(80);
  bigTarget.render();

  Target smallTarget = new Target(20);
  smallTarget.render();
}

class Target
{
  int column;
  int row;
  int gold;// diameter of gold circle

  Target(int givenGold)
  {
    column = int( random(width));
    row = int( random(height));
    gold = givenGold;
  }

  void render()
  {
    fill(0, 158, 206); // blue
    ellipse(column, row, gold * 3, gold * 3);

    fill(217, 25, 25); // red
    ellipse(column, row, gold * 2, gold * 2);

    fill(251, 212, 0); // gold
    ellipse(column, row, gold, gold);
  }
}
```

Index